Lecture Notes in Computer S

Edited by G. Goos and J. Hartmanis

Advisory Board: W. Brauer   D. Gries   J. Stoer

842

Lecture Notes in Computer Science
Edited by G. Goos and J. Hartmanis

Advisory Board: W. Brauer D. Gries J. Stoer

Ton Kloks

Treewidth

Computations and Approximations

Springer-Verlag

**Berlin Heidelberg New York
London Paris Tokyo
Hong Kong Barcelona
Budapest**

Series Editors

Gerhard Goos
Universität Karlsruhe
Postfach 69 80
Vincenz-Priessnitz-Straße 1
D-76131 Karlsruhe, Germany

Juris Hartmanis
Cornell University
Department of Computer Science
4130 Upson Hall
Ithaca, NY 14853, USA

Author

Ton Kloks
Department of Mathematics and Computing Science
Eindhoven University of Technology
P. O. Box 513, 5600 MB Eindhoven, The Netherlands

CR Subject Classification (1991): G.2.2, F.2.2, I.3.5

ISBN 3-540-58356-4 Springer-Verlag Berlin Heidelberg New York
ISBN 0-387-58356-4 Springer-Verlag New York Berlin Heidelberg

CIP data applied for

This work is subject to copyright. All rights are reserved, whether the whole or part of the material is concerned, specifically the rights of translation, reprinting, re-use of illustrations, recitation, broadcasting, reproduction on microfilms or in any other way, and storage in data banks. Duplication of this publication or parts thereof is permitted only under the provisions of the German Copyright Law of September 9, 1965, in its current version, and permission for use must always be obtained from Springer-Verlag. Violations are liable for prosecution under the German Copyright Law.

© Springer-Verlag Berlin Heidelberg 1994
Printed in Germany

Typesetting: Camera-ready by author
SPIN: 10475532   45/3140-543210 - Printed on acid-free paper

Preface

This book deals with problems related to the treewidth parameter of graphs. Tree-decompositions and path-decompositions of graphs of bounded tree- or pathwidth have been found applicable to a wide range of difficult problems. The study of the treewidth parameter has led to a theory with many deep results and challenging open problems.

The book starts with an introduction containing a short overview of the contents of the various chapters. The second chapter offers a smooth and gentle introduction to the most fundamental properties of treewidth, pathwidth and triangulations of graphs. The next three chapters deal with problems that are solved using structural properties of graphs with bounded treewidth. Chapter 6 deals with the approximation of treewidth for graphs in general. Chapter 7 up through 12 illustrate various methods for approximations and computations of treewidth and pathwidth for various classes of graphs. Finally, in Chapter 13, 14 and 15 we describe linear time algorithms for the computation of treewidth and pathwidth of treewidth-bounded graphs and pathwidth-bounded graphs.

The research for this book, which is partly a revised version of my PhD thesis, was primarily done in the period from June 1989 to June 1993 when I was at the department of computer science of Utrecht University (The Netherlands). I would like to thank the people of the computer science department at Utrecht for creating a pleasant and stimulating working atmosphere. Especially, I thank Hans Bodlaender for introducing me to this area of research. I also thank Jan van Leeuwen. This book has benefitted very much from his comments and suggestions. I am grateful to Dieter Kratsch of the department of mathematics and computer science of the Friedrich-Schiller university at Jena (Germany) for drawing my attention to perfect graphs. I thank him and Haiko Müller for inviting me to Jena and for the many discussions on special graph classes we had during that time.

I prepared the final version of this book at the mathematics and computing science department at Eindhoven University. I thank the people of the 'Club' who have read and commented on various parts of the book. Especially I thank Anne Kaldewaij for giving me the opportunity and the stimulation to work on this book from June 1993 to June 1994. Finally, I thank my wife Annelies Jacobs. She took part in everything except in typing the manuscript.

June 1994   Ton Kloks

Contents

Chapter 1

Introduction and basic terminology

In this book we consider a number of problems related to treewidth and path-width of graphs. The main goal is to obtain good bounds on the complexity of determining the treewidth and pathwidth for various interesting classes of graphs. In this section we review some basic concepts and notations and outline the main results of the book.

We assume that the reader is familiar with the most well-known notions from graph theory (see, e.g., [82]). Unless stated otherwise we assume that all graphs are finite, undirected and without loops or multiple edges.

As a standard notation we use $G = (V, E)$, where G is a graph with vertex set V and edge set E. We refer to elements of V as vertices and to elements of E as edges. Each edge is an unordered pair of vertices. Two different vertices are called adjacent if they are both incident with the same edge. For each vertex x we denote by $N(x)$ the neighborhood of x, the set of vertices which are adjacent to x.

Let $G = (V, E)$ be a graph. A subgraph of G is a graph $H = (V', E')$ with $V' \subseteq V$ and $E' \subseteq E$. If $W \subseteq V$ is a subset of vertices then $G[W]$ denotes the subgraph *induced* by W, i.e., $G[W]$ is the subgraph with vertex set W in which two vertices are adjacent whenever they are adjacent in G. If $W = \{v_1, \ldots, v_s\}$ then we sometimes use the notation $G[v_1, \ldots, v_s]$ for $G[W]$. If W is a subset of vertices such that every pair of vertices is adjacent then W is called a clique in G. The maximum number of vertices in a clique in G is called the clique number of G and is denoted by $\omega(G)$.

Let $G = (V, E)$ be a graph. A vertex coloring is a mapping $f : V \to S$ from the vertex set into a set S of colors such that no two adjacent vertices have the same color, i.e., $f(x) \neq f(y)$ whenever x and y are adjacent. The minimum number of colors for which there exists a vertex coloring of a graph G is called the chromatic number, $\chi(G)$.

A walk is an alternating sequence of vertices and edges, starting and ending with a vertex, in which every edge is incident with the two vertices immediately preceding and following it. If all vertices of a walk are distinct (and hence also all edges in the walk are distinct) the walk is called a path. If the walk is closed, i.e., the starting and ending vertex are the same, and if all vertices in the walk are distinct, then the walk is called a cycle. If two vertices of a cycle in G are adjacent if and only if the incident edge is also in the cycle, then the cycle is called chordless. A (chordless) cycle with three vertices is called a triangle and a chordless cycle with four vertices is called a square.

A tree is a special type of graph, which is connected (i.e., every two vertices are connected by a path) and does not contain any cycle. The vertices in a tree usually are called nodes or points. A natural generalization of trees are the so-called k-trees. A k-tree is a graph which can be recursively defined as follows. A clique with $k+1$ vertices is a k-tree and a k-tree with $n+1$ vertices can be obtained from a k-tree with n vertices by making a new vertex adjacent to exactly all vertices of a k-clique. Subgraphs of k-trees are called partial k-trees. If a partial k-tree G is a subgraph of a k-tree H, we call H a k-tree embedding for G.

The minimum value k for which a graph is a subgraph of a k-tree is called the treewidth of a graph. It is not difficult to see that k-trees are partial ℓ-trees for every $\ell \geq k$. Hence every graph with treewidth k is a partial ℓ-tree for every $\ell \geq k$, i.e., the class of partial k-trees is exactly the class of graphs with treewidth at most k.

Because of the underlying tree-like structure, it turns out that for many difficult (e.g., NP-hard) problems, there exist algorithms of which the running time is polynomial in the number of vertices but exponential in the treewidth k of a graph G, if a k-tree embedding of G is given. It follows that for each value k these problems are solvable in polynomial time when restricted to the class of graphs with treewidth at most k, assuming a k-tree embedding of the graph is part of the input. It is an NP-complete problem to decide whether or not for a given graph and given integer k, the graph has treewidth at most k [4]. So finding a k-tree embedding of the input graph for which the value k is as small as possible is a hard problem. Fortunately, in many important special cases this problem can be solved efficiently. For example, we show in chapter 13, chapter 14 and chapter 15 that for each value k there is a linear time algorithm to decide whether a given graph G has treewidth (or pathwidth) k. Furthermore, if the graph has indeed treewidth at most k, the algorithm can be implemented such that it outputs a k-tree of which the graph is a subgraph. As a consequence, we may conclude that for each k there exist fast algorithms for many NP-hard problems when restricted to the class of partial k-trees.

The existence of polynomial time algorithms for treewidth and pathwidth for graphs of bounded treewidth and pathwidth respectively, can also be shown by using the theory of graph minors. Although we do not use much of the theory of graph minors, we include a short list of the most important results. If G is a graph and (u, v) is an edge of G, the graph obtained by *contracting* the edge (u, v) is the graph obtained from $G[V \setminus \{u, v\}]$, by adding a new vertex z and adding edges (z, w) for all $w \in (N(u) \cup N(v)) \setminus \{u, v\}$. A graph H obtained from G by a series of vertex deletions, edge deletions and contractions is called a *minor* of G. A class of graphs $\mathcal{F}$ is called minor-closed if for every graph G in $\mathcal{F}$, every minor of G is also a member of $\mathcal{F}$. For a class of graphs $\mathcal{F}$, a finite obstruction set S is a finite set of minors such that a graph is a member of $\mathcal{F}$ if and only if it does not contain an element of S as a minor. Some of the most important results and conjectures from the theory of graph minors are listed below.

1. Kuratowski showed that a graph is planar if and only if it does not contain K_5 or $K_{3,3}$ as a minor.

2. Hadwiger's conjecture states that if a graph does not have K_s as a minor, then the chromatic number is less than s. Wagner showed that in the case of $s = 5$, this is implied by the 4-Color-Conjecture [169].

3. Wagner's conjecture, stating that every minor-closed class of graphs has a finite obstruction set, was proved by Robertson and Seymour [142].

4. If X is a graph, and $\mathcal{G}[HX]$ the class of graphs with no minor isomorphic to X, then there is a constant c such that all graphs of $\mathcal{G}[HX]$ have treewidth $\leq c$ if and only if X is planar. This was proved by Robertson and Seymour [144].

5. For every graph H, there is an $O(n^3)$ algorithm to test whether H is a minor of a given graph G with n vertices. It follows that the graphs in every minor-closed class of graphs are recognizable in $O(n^3)$ time [142].

6. For every graph H and every constant k, there is an $O(n)$ algorithm to test whether H is a minor of a given graph G with n vertices and treewidth at most k. It follows that every minor-closed class of graphs which does not contain all planar graphs is recognizable in $O(n)$ time (see chapters 13, 14 and 15).

Notice that for each constant k, the class of partial k-trees is minor-closed. For $k = 2, 3$ the obstruction sets have been determined [6]. For larger values of k there does not seem to be a practical way to do this.

We now give a brief overview of the coming chapters. In chapter 2, we start with some basic terminology and tools which are useful when dealing with k-trees and chordal graphs in general. We introduce tree-decompositions of width k of a partial k-tree G as an important alternative and equivalent way to describe k-tree embeddings of G. A special type of tree-decompositions are path-decompositions, and the related embeddings in triangulated graphs are interval graphs.

Chapters 3, 4 and chapter 5 deal with problems related to the structure of partial k-trees. In chapter 3 we show that for each k there exists a linear time algorithm to test superperfection of k-trees. The fact that the class of superperfect graphs is indeed very large is illustrated by the fact that all comparability graphs and hence, for example, all bipartite graphs, are superperfect. Strangely enough, there are only few other classes of graphs for which superperfection can be tested in polynomial time. Indeed we do not even know if the problem is in NP. We show that for each k, there is a finite list of structures, which can be computed in $O(1)$ time, such that a k-tree is superperfect if and only if it does not contain a structure from this list. It is then easy to see that for each k testing superperfection of k-trees can be done in linear time.

In chapter 4 a complete characterization is given for the structure of partial 2-trees. With this characterization we are able to solve the following problem in linear time and using a linear amount of memory: Given a graph of which each vertex is colored with one of three colors such that adjacent vertices have a different color, find, if it exists, an embedding of the graph in a triangulated graph which is also properly colored. If such an embedding exists then there also exists an embedding in a 2-tree. The general problem, to find a k-tree embedding which is properly colored for a graph colored with k colors, stems from molecular biology where it is known as the 'perfect phylogeny problem'.

In chapter 5 we show that a random graph with a number of edges which is only linear in the number of vertices, has, with high probability, a large treewidth (proportional to the number of vertices). Perhaps this illustrates the importance to restrict the search for graph with bounded treewidth to graphs which already have some underlying structure. Research in this direction is reported in chapter 7, 8, 9, 10 and chapter 12.

First, in chapter 6 we look at the problem to find approximate tree-decompositions and path-decompositions for graphs in general. We show that the problem is closely related to that of finding balanced vertex separators in graphs. We use two of these algorithms.

The first is an algorithm of Alon et al. [1] that finds a good vertex separator with approximation ratio $O(\sqrt{kn})$. Using this algorithm as a basic tool we show there exists a fast algorithm that finds a path-decomposition of width $O(k^{3/2}\sqrt{n})$, where k is the treewidth of the graph. An interesting side effect of this algorithm is that it computes, for a *planar* graph, a path-decomposition of width $O(\sqrt{n})$.

An approximation for vertex separators within a factor $O(\log n)$ due to Leighton and Rao was reported in [95]. We use this to develop an algorithm which finds a tree-decomposition of width within a factor $O(\log n)$ off the optimum.

Since the algorithms to compute a tree-decomposition of a graph with treewidth at most k are of practical use only for small values of k at the moment, it is of great importance to devise fast algorithms for determining the treewidth for classes of graphs which are of special practical importance. For some classes of graphs the treewidth and pathwidth can be computed in polynomial time. For example this is the case for cographs [32]. The treewidth can also be computed efficiently for circular arc graphs [159].

In chapter 7, we show that for many important classes of perfect graphs, there exist very fast algorithms which approximate treewidth within a constant factor. For example we give algorithms to approximate treewidth within a constant factor for cotriangulated graphs, split graphs, convex graphs and permutation graphs. For the class of cotriangulated graphs we also give exact algorithms to compute the pathwidth and treewidth. Recently, for the special case of complements of interval graphs, it was shown that the treewidth and pathwidth are equal and can be determined in linear time [69].

For the large class of cocomparability graphs, we give a fast approximation algorithm to compute a tree-decomposition of width at most k^2 where k is the treewidth of the graph. As an important, and perhaps surprising side-effect, we see that for all these graphs the approximate tree-decomposition we construct is in fact a path-decomposition, which shows a strong relationship between the pathwidth and treewidth of these graphs.

In chapter 8 we give a polynomial time algorithm to compute the treewidth of chordal bipartite graphs. This is an important subclass of the class of bipartite graphs, containing many other interesting classes of bipartite graphs. Our result is even more of interest since the problem of determining the treewidth of bipartite graphs in general is NP-hard [4].

In chapter 9 we give a fast algorithm to compute the treewidth and pathwidth of a permutation graph. We show that this generalizes in the following way. For each d there is a polynomial time algorithm to compute the treewidth of a cocomparability graph of which the partial order dimension of the complement is bounded by d. This is of special interest since treewidth is NP-hard for cocomparability graphs in general. Here we assume that a concatenation of a minimal number of permutation diagrams is given as an intersection model. The existence of such a model is guaranteed (see [77]).

It turns out that the treewidth and pathwidth of a permutation graph are equal, and in fact this is true for all cocomparability graphs. This was generalized recently by M. Habib and R. Möhring [79] who showed that for all AT-free graphs the treewidth and pathwidth are equal.

In chapter 10 we generalize some of the results for permutation graphs to the class of circle graphs. We show that there exists an $O(n^3)$ algorithm to determine the treewidth of a circle graph.

It becomes clear in chapters 8, 9 and 10 that minimal separators play a crucial role in determining treewidth (and pathwidth) of special graph

classes. In fact, the graph classes treated in these chapters all have a polynomial number of minimal separators (polynomial in the number of vertices of
the graph). Many other special graph classes have this property. For example this is the case for distance hereditary graphs, circular arc graphs, weakly
triangulated graphs etc. We show in chapter 11 an algorithm to compute all
minimal separators of a graph. The algorithm needs time polynomial in the
number of minimal separators.

We illustrate how minimal separators can be used to find the treewidth
and pathwidth of graphs in chapter 12. Here we give another polynomial
time algorithm to compute the treewidth (and pathwidth) of cocomparability
graphs of bounded dimension. The algorithm of chapter 9 explicitly uses an
intersection model. An important property of the algorithm presented in
chapter 12 is that it does not use an intersection model. This is important,
since finding a suitable intersection model is an NP-hard problem.

In chapter 13 and chapter 14 we show that for each k the decision problems 'treewidth $\leq k$' and 'pathwidth $\leq k$' can be solved in $O(n)$ time when a
tree-decomposition of the graph with bounded width is given. Using a result
of [139], stating that for every constant k there exists an $O(n \log(n))$ algorithm that given a graph with n vertices either outputs that the treewidth
is larger than k, or outputs a tree-decomposition of width at most $3k + 2$
(see also [20]), we obtain, for each constant k, $O(n \log(n))$ algorithms to
decide whether the treewidth or pathwidth is at most k. Furthermore, a
path- or tree-decomposition of optimal width can be found within the same
time bound.

Recently, using the results of chapter 14, it was shown in [21] that for each
constant k it can be decided in linear time whether a graph has treewidth
at most k. We describe this algorithm in chapter 15. Using this result, the
algorithm of chapter 13 gives, for every constant k, a linear time algorithm
to determine if the pathwidth of a graph is at most k. This can be seen as
follows. If the pathwidth of a graph is at most k, then also the treewidth
is at most k. The algorithm of chapther 15 can be used to find a tree-
decomposition with width at most k. Using this tree-decomposition, the
algorithm of chapter 13 can be used to decide if the pathwidth of the graph
is at most k. If this is the case, a path-decomposition with width at most k
can also be found in linear time.

We like to mention here a surprising new result obtained by an improve-
ment of the algorithm described in chapter 13 (see [29]). The pathwidth can
be computed in polynomial time for graphs with bounded treewidth.

Chapter 2

Preliminaries

We review some basic terminology and aspects of triangulated graphs. We define the concepts treewidth and pathwidth of a graph in terms of subgraphs of triangulated graphs. We then show the equivalence between the treewidth and pathwidth of a graph and the minimum width of a tree-decomposition and path-decomposition. We end this section with some notes on perfect graphs.

2.1 Triangulations

In this section we review some important aspects of triangulated graphs and of triangulations of graphs. The various names given to triangulated graphs, such as chordal graphs, rigid circuit graphs, perfect elimination graphs, monotone transitive graphs and so on, illustrate the importance of this class of graphs. We shall mainly use the term triangulated graphs. For a good overview of the different aspects of triangulated graphs the reader is referred to Golumbic's book [76].

Definition 2.1.1 *A graph is* triangulated *if it contains no chordless cycle of length greater than three.*

This is equivalent to saying that the graph does not contain an induced subgraph isomorphic to C_n (i.e., a cycle of length n) for $n > 3$. Notice that being triangulated is a hereditary property, i.e., if a graph G is triangulated then every induced subgraph of G also is triangulated.

There are many ways to characterize triangulated graphs. Characterizations have been appearing from 1960 [56] until very recently [11]. Although many of these characterizations are interesting and useful, it suffices for our purposes to list only some of them.

One of the most important tools that can be used when working with triangulated graphs is the concept of a perfect elimination scheme.

Definition 2.1.2 *Let* $G = (V, E)$ *be a graph. A* simplicial vertex *of* G *is a vertex of which the neighborhood induces a clique. An ordering of the vertices* $\sigma = [v_1, \ldots, v_n]$ *is called a* perfect elimination scheme *if for every* $1 \leq i \leq n$, v_i *is a simplicial vertex in* $G[v_i, \ldots, v_n]$.

Fulkerson and Gross, in a paper in 1965 [66], characterized triangulated graphs by means of a perfect elimination scheme.

Lemma 2.1.1 *A graph* G *is triangulated if and only if there exists a perfect elimination scheme for* G. *Moreover, if a graph is triangulated, any simplicial vertex can start a perfect elimination scheme for it.*

In 1975 Rose and Tarjan [148] gave a linear time algorithm for recognizing triangulated graphs, by finding a perfect elimination scheme if it exists. Many NP-hard problems are solvable in linear time when restricted to triangulated graphs like CLIQUE, INDEPENDENT SET, CLIQUE COVER etc. Lemma 2.1.1 has many useful consequences. For example, every triangulated graph has at least one simplicial vertex. In fact, Dirac showed that if a graph is not a clique, then it has at least two nonadjacent simplicial vertices. This can be used to obtain yet another characterization of triangulated graphs: a graph is triangulated if and only if every induced subgraph is either a clique or has two nonadjacent simplicial vertices. We conclude that if G is triangulated and C is the vertex set of a maximal clique in G, then there exists a perfect elimination scheme $\sigma = [v_1, \ldots, v_n]$ such that $C = \{v_{n-t+1}, \ldots, v_n\}$ (where $t = |C|$).

Notice that if x is a simplicial vertex in a triangulated graph, then $\{x\} \cup N(x)$ is a maximal clique, and in fact this is the only maximal clique containing x. It follows that there are at most n maximal cliques, with equality holding if and only if the graph has no edges. This was first pointed out in [66]. Notice that if $\sigma = [v_1, \ldots, v_n]$ is a perfect elimination scheme, then all maximal cliques are of the form $\{v_i\} \cup (N(v_i) \cap \{v_i, \ldots, v_n\})$.

The last immediate consequence of Lemma 2.1.1 that we mention here is that a triangulated graph is perfect (a graph is called perfect if the clique number is equal to the chromatic number for every induced subgraph; see section 2.3). This can be seen as follows. First, since every induced subgraph of a triangulated graph is again triangulated, we only have to show that the chromatic number $\chi(G)$ and clique number $\omega(G)$ are equal for any triangulated graph G. To prove it, let $\sigma = [v_1, \ldots, v_n]$ be a perfect elimination scheme. For $i = n, n-1, \ldots, 1$, color v_i with a color which is not used in $N(v_i) \cap \{v_{i+1}, \ldots, v_n\}$. This gives a correct coloring, and since each $\{v_i\} \cup (N(v_i) \cap \{v_{i+1}, \ldots, v_n\})$ is a clique, this coloring can be done using no more than $\omega(G)$ colors.

Another characterization of triangulated graphs was given by Dirac in 1961 [56], by means of minimal vertex separators.

Definition 2.1.3 *Given a graph* G *with vertex set* V, *a subset* S ⊂ V *is called a* vertex separator *for nonadjacent vertices* a *and* b *in* V \ S *if* a *and* b *are in different connected components of* G[V \ S]. *If* S *is a vertex separator for* a *and* b *but no proper subset of* S *separates* a *and* b *in this way, then* S *is called a* minimal vertex separator *for* a *and* b *or a minimal* a, b-separator. *A subset* S ⊆ V *is called a* minimal vertex separator *if there exists a pair of nonadjacent vertices for which* S *is a minimal vertex separator.*

Since we deal only with vertex separators, we usually call a (minimal) vertex separator simply a (minimal) separator. We like to stress that one minimal vertex separator might well be properly contained in another minimal vertex separator.

Lemma 2.1.2 *A graph* G *is triangulated if and only if every minimal vertex separator induces a complete subgraph of* G.

The following lemma, which must have been rediscovered many times, appears as an exercise in [76].

Lemma 2.1.3 *Let* S *be a minimal vertex separator for nonadjacent vertices* a *and* b, *and let* C_a *and* C_b *be the connected components of* G[V\S] *containing* a *and* b, *respectively. Then every vertex of* S *has a neighbor in* C_a *and a neighbor in* C_b.

Proof. Assume $w \in S$ has no neighbor in C_a. We claim that $S' = S \setminus \{w\}$ is also an a, b-separator. Suppose there is a path between a and b in G[V\S']. Then the path contains no vertex of S', hence the path passes through w. This leads to a contradiction.   □

Especially useful is the fact that a triangulated graph has *balanced* separators. The following lemma is a generalization of probably the oldest separator theorem of which one version is due to C. Jordan in 1869 [90, 111]. For a similar result see e.g. [143].

Lemma 2.1.4 *Let* G = (V, E) *be a triangulated graph with* n *vertices. There is a clique* C, *such that every component of* G[V \ C] *has at most* $\lceil \frac{1}{2}(n - |C|) \rceil$ *vertices.*

As mentioned earlier, there are many more characterizations of triangulated graphs. For example, a graph is triangulated if and only if every induced connected subgraph with $p \geq 2$ vertices contains at most $p - 1$ maximal cliques (see for example [15] page 84). Another interesting characterization appears in [11] and uses the so-called stability function of a graph. Let G be a graph with vertex set $\{v_1, \ldots, v_n\}$. With each vertex we can associate a 0/1-variable x_i. If x is some vector in $\{0, 1\}^n$ then the subgraph of

G induced by x is defined as the subgraph induced by exactly those vertices v_i for which the corresponding variable x_i is set to one. The stability function $\alpha_G : \{0,1\}^n \to \mathbf{N}$ is defined as follows. For each $x \in \{0,1\}^n$, $\alpha_G(x)$ is the stablity number (the maximum cardinality of an independent set) of the subgraph induced by x. It can be shown that this function can be expressed uniquely in the form $\sum_{t \in \Delta} a_t \prod_{i \in t} x_i$, where Δ is a collection of subsets of $\{1, \ldots, n\}$ and the a_t's are real coefficients. Now the following statement holds. A graph is triangulated if and only if the polynomial expression of the stability function has all its coefficients in $\{0, -1, 1\}$.

We conclude with one more characterization, showing that triangulated graphs are a class of intersection graphs. This was shown for example in [172].

Lemma 2.1.5 *A graph* G *is triangulated if and only if* G *is the intersection graph of a family of subtrees of a tree.*

The kind of intersection graph referred to in the lemma is defined as follows. Given a family of subtrees of a tree, a graph is constructed in the following way. The vertices of the graph are the subtrees and two vertices are adjacent if the corresponding subtrees have at least one node in common. Notice that a family of subtrees of a tree satisfies the Helly property, which is the following (see for example [76]).

Definition 2.1.4 *A family* $\{S_i\}_{i \in I}$ *of sets satisfies the* Helly *property if for all* $J \subset I$: $S_i \cap S_j \neq \emptyset$ *for all* $i, j \in J$ *implies that* $\cap_{j \in J} S_j \neq \emptyset$.

Notice that this implies the following. Assume we have a set of connected subgraphs of a triangulated graph such that for every pair the subgraphs have at least one vertex in common. Then there is at least one vertex present in each of the subgraphs.

An important subclass of the triangulated graphs are the interval graphs (which were first mentioned by Hajös [80]).

Definition 2.1.5 *An* interval graph *is a graph for which one can associate with each vertex an interval on the real line such that two vertices are adjacent if and only if their corresponding intervals have a nonempty intersection.*

Definition 2.1.6 *An ordering* $(X_1, \ldots, X_t)$ *of the maximal cliques of a graph* G *is called an* interval ordering *if, for every vertex, the maximal cliques containing it occur consecutively in the ordering.*

An interval ordering of the maximal cliques of a graph is sometimes called a *consecutive clique arrangement.*

The following characterization of interval graphs was found by Gilmore and Hoffman in 1964 [72].

Lemma 2.1.6 *The following statements are equivalent.*

1. *G is an interval graph.*

2. *G is triangulated and its complement $\overline{G}$ is a comparability graph.*

3. *There is an interval ordering of the maximal cliques of G.*

A comparability graph is a graph which admits a transitive orientation of its edges [76].

Interval graphs have many practical applications in various fields like archeology, geology, criminology, genetics etc., see for example [46, 76]. For a much more extensive bibliographical list of applications we refer to page 90 of the overview paper of Duchet which appeared in [15].

As an interval graph is triangulated, the number of maximal cliques is bounded by the number of vertices. By Lemma 2.1.6 the maximal clique-versus-vertex incidence matrix has the consecutive ones property. In [35] Booth and Lueker give a fast algorithm to test for the consecutive ones property using PQ-trees, which implies the following result:

Lemma 2.1.7 *A graph $G = (V, E)$ can be tested for being an interval graph in $O(n + e)$ steps, where n is the number of vertices and e is the number of edges of G.*

Recently a simpler algorithm was discovered by Hsu [88] which does not use the maximal cliques and places the intervals directly. As with triangulated graphs, there are many more characterizations for interval graphs. For example, Lekkerkerker and Boland in 1962, found a list of graphs such that a graph is an interval graph if and only if it does not contain an induced subgraph isomorphic to a graph in this list [120]. Since the complements of interval graphs are comparability graphs, such a list is also easily obtained from the complete list of forbidden induced subgraphs for comparability graphs, which was found by Gallai [68] in 1967 (this list appears also in [15] on page 78). Lekkerkerker and Boland also proved another important characterization, using astroidal triples.

Definition 2.1.7 *Three vertices in a graph G are called an* astroidal triple *if any two of them is connected by a path which avoids the neighborhood of the third.*

For example, in a 3-sun (see figure 3.1 on page 30) the three vertices of the independent set are an astroidal triple.

Lemma 2.1.8 *A graph is an interval graph if and only if G is triangulated and does not contain an astroidal triple.*

A second important subclass of triangulated graphs is the class of k-trees. The first mention we found of 2-trees is in a paper of Harary and Palmer [83] and dates back to 1968 (in this paper they are introduced as being simple connected, acyclic 2-plexes). The related concept of a (acyclic) pure k-complex is defined in a paper of Harary in 1955 [81]. When restricted to 2-trees another related concept, called a two-terminal series-parallel network, can even be traced back to Macmahon in 1892 [123], who described a method to determine the number of distinct series-parallel two-terminal networks. This is mentioned in a paper of R. M. Foster [64] (see also [65]). Foster counts the number of different series-parallel networks without specification of terminals. (He mentions that in [163] the enumeration through $n = 7$ has been performed by Tellegen.)

In a paper from 1969 Beineke and Pippert enumerate labeled k-dimensional trees (after enumerating 2-dimensional trees in 1968) which they call k-trees for short [10].

Usually a clique with k vertices is also considered to be a k-tree. For convenience we prefer to define k-trees to have at least $k + 1$ vertices. With this definition, the treewidth of a k-tree is k (the treewidth of a k-clique is $k - 1$), and the clique number of a k-tree is $k + 1$.

Definition 2.1.8 k-Trees *are defined recursively as follows: A clique with $k + 1$ vertices is a k-tree; given a k-tree T_n with n vertices, a k-tree with $n+1$ vertices is constructed by taking T_n and creating a new vertex x_{n+1} which is made adjacent to a k-clique of T_n and nonadjacent to the $n - k$ other vertices of T_n.*

The first characterization shows that k-trees are indeed triangulated.

Lemma 2.1.9 *A graph G with n vertices is a k-tree if and only if $n > k$ and there exists a perfect elimination scheme $\sigma = [v_1, \ldots, v_n]$ such that for all $i \le n-k$, v_i is adjacent to a k-clique in the subgraph $G[v_i, \ldots, v_n]$.*

Proof. Assume $G = (V, E)$ is a k-tree with n vertices. If $n = k+1$ then G is a $(k + 1)$-clique and any ordering of the vertices yields a perfect elimination scheme satisfying the desired property. If $n > k + 1$, by definition there is a vertex x such that $G[V \setminus \{x\}]$ is a k-tree with $n - 1$ vertices, and x is adjacent to a k-clique. By induction we may assume that there is a perfect elimination scheme $\sigma' = [v_2, \ldots, v_n]$ for $G[V \setminus \{x\}]$ such that v_i is adjacent to a k-clique in $G[\{v_i, \ldots, v_n\}]$, for all $2 \le i \le n - k$. Define $v_1 = x$. Then $\sigma = [v_1, \ldots, v_n]$ is a perfect elimination scheme as mentioned for G.

Now assume $n > k$ and $\sigma = [v_1, \ldots, v_n]$ is a perfect elimination scheme such that for all $i \le n - k$, v_i is adjacent to a k-clique in $G[v_i, \ldots, v_n]$. If $n = k + 1$ it follows that the graph must be a $(k + 1)$-clique and hence a k-tree. Assume $n > k + 1$. Then $\sigma' = [v_2, \ldots, v_n]$ is a perfect elimination scheme for $G[V \setminus \{v_1\}]$ with the stated property guaranteeing by induction that $G[V \setminus \{v_1\}]$ is a k-tree. Since v_1 is adjacent to a k-clique it follows by definition of a k-tree that G is a k-tree. □

It follows that all maximal cliques in a k-tree have $k + 1$ vertices. Lemma 2.1.9 makes it very easy to calculate the number of certain subgraphs in a k-tree. We give one example.

Lemma 2.1.10 *For* $1 \leq t \leq k + 1$, *the number of t-cliques in a k-tree is*

$$\binom{k}{t} + (n - k)\binom{k}{t-1}$$

The following characterization of k-trees appears in [147].

Lemma 2.1.11 *A graph* G *is a k-tree if and only if*

1. G *is connected,*

2. $\omega(G) = k + 1$,

3. *every minimal vertex separator of* G *is a k-clique.*

The following lemma shows that k-trees are triangulated graphs with clique number $k + 1$ which have a maximum number of edges (see also [147]). In Lemma 2.1.13 we will show that, conversely, every triangulated graph with clique number $k + 1$ and with a maximal number of edges is a k-tree.

Lemma 2.1.12 *A graph* G *with* n *vertices is a k-tree if and only if*

1. G *is triangulated,*

2. $\omega(G) = k + 1$,

3. *the number of edges is at least* $nk - \frac{1}{2}k(k + 1)$.

Proof. First assume G is a k-tree. By Lemma 2.1.9 G is triangulated and has clique number $k + 1$. By Lemma 2.1.10 the number of edges in a k-tree is $nk - \frac{1}{2}k(k + 1)$. Conversely, assume that a graph G satisfies the three conditions. Since G is triangulated there is a perfect elimination scheme $\sigma = [v_1, \ldots, v_n]$. Since $\omega(G) = k + 1$, it follows that for each i, $|N(v_i) \cap \{v_i, \ldots, v_n\}| \leq k$. Hence we find for the number of edges in G:

$$\sum_{i=1}^{n} |N(v_i) \cap \{v_i, \ldots, v_n\}| \leq \sum_{i=1}^{n-k} k + \frac{1}{2}k(k - 1) = nk - \frac{1}{2}k(k + 1)$$

We can have equality only if for each $i \leq n - k$ v_i has k neighbors in $\{v_i, \ldots, v_n\}$. By Lemma 2.1.9 G is a k-tree. □

Definition 2.1.9 *A* partial k-tree *is any subgraph of a k-tree.*

There exist linear time algorithms for many NP-complete problems when
these problems are restricted to the class of partial k-trees for some *constant*
k and when an embedding in a k-tree (i.e., a k-tree of which the graph is
a subgraph) is given [3, 5, 7, 18]. There are also results stating that large
classes of problems can be solved in linear time for the class of partial k-trees
with k bounded by some constant [48, 49, 50].

In this book we shall primarily be concerned with the problem of finding
nice embeddings of a graph in triangulated graphs and in interval graphs.
We call such an embedding a triangulation of the graph.

Definition 2.1.10 *A triangulation of a graph* G *is a graph* H *with the
same set of vertices such that* G *is a subgraph of* H *and such that* H *is
triangulated. We say that* G *is* triangulated into H.

Clearly, every graph can be triangulated (into a clique). There are two
problems which have drawn much attention because of the large number
of applications. The first problem is to triangulate a graph such that the
number of added edges is minimum and the second one is to triangulate a
graph such that the clique number in the triangulated graph is minimum.
The first one is called the MINIMUM FILL-IN problem and the second one
is the TREEWIDTH problem. These problems are both NP-hard [4, 174].
Our concern, as one might guess by the title of this book, lies in finding
embeddings which minimize the maximum clique.

Remark. The fact that these problems are indeed different can be seen by
the following example [24]. Let $\delta > 3$ be an integer. Let C be a cycle with
$2\delta + 3$ vertices, I an independent set with δ vertices and K a clique with
2δ vertices. Consider the graph $G = C + (I \cup K)$. There are two possible
triangulations with a minimal number of edges. In the first edges are added
such that $I \cup K$ becomes a clique and such that C becomes triangulated. In
this case the clique number becomes $3\delta + 3$ and the number of added edges
is $\frac{1}{2}\delta(5\delta + 3)$. The second possible triangulation with a minimal number of
edges, adds edges such that C becomes a clique. In this way the maximum
clique has $4\delta + 3$ vertices and the number of added edges is $\delta(2\delta + 3)$. Hence
the first triangulation minimizes the clique number and the second minimizes
the number of edges.

Another result obtained in [24] is the following. An elimination scheme
for a graph G is simply an ordering of the vertices $\sigma = [v_1, \ldots, v_n]$. Given
an elimination ordering σ let $G' = G(\sigma)$ be the graph obtained from G by
adding the minimum number of edges such that σ is a perfect elimination
scheme for G'. Let a graph G have property $\mathcal{P}$ if for *every* elimination
ordering σ the triangulation $G(\sigma)$ realizes the minimum fill-in. In [24] it is
shown that the only graphs having property $\mathcal{P}$ are graph for which every

connected component is either a clique, a cycle or a cocktailparty graph (i.e. the complement of $K_2 \cup K_2 \cup \ldots \cup K_2$). For these graphs every $G(\sigma)$ also minimizes the clique number. As far as we know, it is an open problem to characterize those graphs for which every triangulation with a minimal number of edges is a triangulation which minimizes the clique number.

Lemma 2.1.13 *If* G *is a triangulated graph with at least* $k + 1$ *vertices and has clique number at most* $k + 1$, *then* G *can be triangulated into a* k*-tree.*

Proof. Let $\sigma = [v_1, \ldots, v_n]$ be a perfect elimination scheme for G. We make an embedding of G in a k-tree recursively as follows. First we add edges such that the subgraph induced by $\{v_{n-k}, v_{n-k+1}, \ldots, v_n\}$ becomes a $k+1$-clique. For the induction step, assume the subgraph with vertices $\{v_{i+1}, \ldots, v_n\}$ has been triangulated into a k-tree T_{i+1}. In G vertex v_i is adjacent to a clique C with at most k vertices in $\{v_{i+1}, \ldots, v_n\}$. In the k-tree T_{i+1} C is contained in a k-clique C'. We make v_i adjacent to all vertices of C' and we obtain a k-tree T_i which is a triangulation of G. □

Lemma 2.1.14 *Every partial* k*-tree with at least* $k + 1$ *vertices can be triangulated into a* k*-tree.*

Proof. Let the partial k-tree $G = (V, E)$ be a subgraph of the k-tree $H = (W, F)$. Then $H[V]$ is triangulated and has clique number at most $k + 1$. By Lemma 2.1.13, $H[V]$ can be triangulated into a k-tree. Clearly this is a triangulation of G. □

We have seen that k-trees are triangulations with a maximal number of edges. We now show some properties of triangulations with a minimal number of edges.

Definition 2.1.11 *A minimal triangulation* H *of a graph* G *is a triangulation of* G *such that the following conditions are satisfied:*

1. *If* a *and* b *are nonadjacent vertices in* H *then every minimal* a, b-*separator of* H *is also a minimal* a, b-*separator in* G,

2. *If* S *is a minimal separator in* H *and* C *is the vertex set of a connected component of* $H[V \setminus S]$, *then* C *induces also a connected component in* $G[V \setminus S]$.

Theorem 2.1.1 *Let* H *be a triangulation of* G. *There exists a minimal triangulation* H' *of* G *such that* H' *is a subgraph of* H.

Proof. Suppose H has a minimal vertex separator W for nonadjacent vertices a and b, such that either W induces no minimal vertex separator for a and b in G, or the vertex sets of the connected components of $H[V \setminus W]$ are different from those of $G[V \setminus W]$. Let $S \subseteq W$ be a minimal a, b-separator in G. Let $C_1, \ldots, C_t$ be the connected components of $G[V \setminus S]$. Make a chordal graph H' as follows. For each $C_i \cup S$ take the chordal subgraph of H induced by these vertices. Since S is a clique in H, this gives a chordal subgraph H' of H. Notice that the vertex sets of the connected components of $H'[V \setminus S]$ are the same as those of $G[V \setminus S]$. We claim that the number of edges of H' is less than the number of edges of H which, by induction, proves the theorem. Clearly H' is a subgraph of H.

First assume that $S \neq W$, and let $x \in W \setminus S$. By Lemma 2.1.3, in H, x has a neighbor in the component containing a and a neighbor in the component containing b. Not both these edges can be present in H'.

Now assume $S = W$. Then, by assumption, the vertex sets of the connected components of $H[V \setminus W]$ are different from those of $H'[V \setminus W]$. Since H' is a subgraph of H, every connected component $H'[V \setminus W]$ is contained in some connected component of $H[V \setminus W]$. It follows that there must be a connected component in $H[V \setminus W]$ containing two different connected components of $H'[V \setminus W]$. This can only be the case if there is some edge between these components in $H[V \setminus W]$ (which is not there in $H'[V \setminus W]$). This proves the theorem.   □

To illustrate that minimal triangulations are not very restrictive, notice that a clique is a minimal triangulation of G. We now show that we can restrict the triangulations to be considered somewhat more.

Definition 2.1.12 *Let Δ be the set of all minimal separators of a graph $G = (V, E)$. For a subset $C \subseteq \Delta$ let G_C be the graph obtained from G by adding edges between vertices contained in the same set $C \in \mathcal{C}$. If the graph G_C is a minimal triangulation of G such that $\mathcal{C}$ is exactly the set of all minimal separators of G_C, then G_C is called an* efficient *triangulation.*

Notice that for each $C \in \mathcal{C}$, the induced subgraph $G_C[C]$ is a clique.

Theorem 2.1.2 *Let H be a triangulation of a graph G. There exists an efficient triangulation G_C of G which is a subgraph of H.*

Proof. Take a minimal triangulation H' which is a subgraph of H such that the number of edges of H' is minimal (theorem 13.3.3). We claim that H' is efficient. Let $\mathcal{C}$ be the set of minimal vertex separators of H'. We prove that $G_C = H'$.

Since every minimal separator in a triangulated graph is a clique, it follows that G_C is a subgraph of H'. Consider a pair of vertices a and b which are adjacent in H′ but not adjacent in G. Remove the edge from the graph H′. Call the resulting graph H*. Since the number of edges of H′ is minimal, it follows that H* has a chordless cycle. Clearly this cycle must have length four. Let $\{x, y, a, b\}$ be the vertices of this square. Then x and y are non adjacent in H′. But then a and b are contained in every minimal x, y-separator in H′. It follows that a and b are also adjacent in G_c.   □

Corollary 2.1.1 *A triangulation of a graph with a minimal number of edges is efficient.*

Let $G = (V, E)$ be a graph and let C be a minimal vertex separator. Let $C_1, \ldots, C_t$ be the connected components of $G[V \setminus C]$. Denote by $\overline{C}_i$ $(i = 1, \ldots, t)$ the graph obtained as follows. Take the induced subgraph $G[C \cup C_i]$, and add all possible edges between vertices of C such that the subgraph induced by C is complete. The following lemma easily follows from Theorem 2.1.1 (a similar result appears in [4]).

Lemma 2.1.15 *A graph G with at least $k + 2$ vertices is a partial k-tree if and only if there exists a minimal vertex separator with at most k vertices such that all graphs $\overline{C}_i$ are partial k-trees.*

Proof. Assume G is a partial k-tree. Let H be a minimal triangulation. Then $\omega(H) \leq k + 1$. Thus, because H has more than $k + 1$ vertices, there must exist nonadjacent vertices in H. Let C be a minimal vertex separator in H. Then this is also a minimal vertex separator in G. Since H is triangulated, C is a clique in H. Let $C_1, \ldots, C_t$ be the vertex sets of the connected components of $H[V \setminus C]$. These are also the vertex sets of the connected components of $G[V \setminus C]$. The graphs $H[C_i \cup C]$ are triangulations of $\overline{C}_i$, hence these are partial k-trees.

Conversely, let C be a minimal vertex separator and let $C_1, \ldots, C_t$ be the vertex sets of the connected components of $G[V \setminus C]$. Assume all $\overline{C}_i$ are partial k-trees. Make triangulations H_i of each of these graphs and identify the vertices of C. We claim that the result is a triangulation H of G. This can be seen for example as follows. Let the vertices of C be $c_1, \ldots, c_w$. For each triangulation H_i take a perfect elimination scheme $\sigma_i = [v_1^i, \ldots, v_{n_i}^i, c_1, \ldots, c_w]$. We can construct a perfect elimination scheme for H of the following form: $\sigma = [v_1^1, \ldots, v_{n_1}^1, \ldots, v_1^t, \ldots, v_{n_t}^t, c_1, \ldots, c_w]$.   □

2.2 Treewidth and pathwidth

In this section we give a brief introduction to some important notions related to partial k-trees: treewidth, pathwidth, tree-decomposition and path-decomposition.

We define the treewidth of a graph in two ways and show the equivalence of the two definitions (see also [20, 117]). The first definition, which we have seen already in the previous section, is by means of partial k-trees and the second is by means of tree-decompositions.

Definition 2.2.1 *The* treewidth *of a graph* G *is the minimum value* k *for which* G *is a partial* k-*tree.*

Determining the treewidth or the pathwidth of a graph is NP-hard [4]. However, for *constant* k, graphs with treewidth $\leq$ k are recognizable in linear time, see chapter 13 and chapter 14.

Notice that a k-clique has treewidth $k - 1$ (since it is a $(k - 1)$-tree). The following result may serve as an example, and is needed in later chapters (see also, e.g., [32]).

Lemma 2.2.1 *The complete bipartite graph* $G = K(m, n)$ *has treewidth equal to* $\min(m, n)$.

Proof. First notice that in *any* triangulation of G at least one of the color classes is a clique. Otherwise there are nonadjacent vertices x and y in one colorclass and nonadjacent vertices p and q in the other colorclass, and $H[x, y, p, q]$ would be a chordless cycle. This shows that the treewidth is at least $\min(m, n)$.

For a bipartite graph $G = (X, Y, E)$ the graph $\text{split}(G) = (X, Y, \hat{E})$ is defined as the graph with $\hat{E} = E \cup \{(x, x') \mid x, x' \in X\}$. A *split graph* is a graph for which there exists a partition of the vertex set into a clique and a stable set. Assume without loss of generality that $|X| \leq |Y|$. If G is complete bipartite then $\text{split}(G)$ is a k-tree with $k = |X|$ consisting of a maximal clique with $|X| + 1$ vertices and a set of $|Y| - 1$ simplicial vertices. This proves that the treewidth is at most $\min(m, n)$. $\square$

We now come to the second way to define the treewidth of a graph, namely by a concept called the tree-decomposition of a graph.

Definition 2.2.2 *A* tree-decomposition *of a graph* $G = (V, E)$ *is a pair* $D = (S, T)$ *with* $S = \{X_i \mid i \in I\}$ *a collection of subsets of vertices of* G *and* $T = (I, F)$ *a tree, with one node for each subset of* S, *such that the following three conditions are satisfied:*

1. $\bigcup_{i \in I} X_i = V$,

2. *for all edges* $(v, w) \in E$ *there is a subset* $X_i \in S$ *such that both* v *and* w *are contained in* X_i,

3. *for each vertex* x *the set of nodes* $\{i \mid x \in X_i\}$ *forms a subtree of* T.

Definition 2.2.3 *A path-decomposition of a graph G is a tree-decomposition* (S, T) *such that* T *is a path. A path-decomposition is also denoted as* $(X_1, X_2, \ldots, X_\ell)$.

An alternative way to formulate the third condition is the following: for all $i, j, k \in I$: if j is on the path from i to k in T then $X_i \cap X_k \subset X_j$. Notice that if D is a tree-decomposition for a graph G and H is a subgraph of G with the same vertex set, then D is also a tree-decomposition for H. Also, if H is a subgraph of G and $D = (S, T)$ is a tree-decomposition for G, we can obtain a tree-decomposition D' for H by taking the restriction of every subset in S to the vertex set of H (the three conditions in Definition 2.2.2 are trivially satisfied).

Definition 2.2.4 *Let* $D = (S, T)$ *be a tree- or path-decomposition for a graph G and let H be any subgraph of G. The tree- or path-decomposition for H obtained from D by taking the restriction of every subset in S to the vertex set of H is called the* subdecomposition *of D for H.*

Definition 2.2.5 *The* width *of a tree-decomposition* $(\{X_i | i \in I\}, T = (I, F))$ *is* $\max_{i \in I}(|X_i| - 1)$.

To show the equivalence between the treewidth of a graph and the minimum width over all tree-decompositions, we make use of the following lemma [32], which is a direct consequence of the Helly property satisfied by a set of subtrees of a tree (for alternative proofs see [18, 150]).

Lemma 2.2.2 *Let* $(S = \{X_i \mid i \in I\}, T = (I, F))$ *be a tree-decomposition for G. For every clique C in G there exists a subset* $X_i \in S$ *such that* $C \subset X_i$.

Proof. Let C be a clique in G. For every pair of vertices x and y of C there exists a subset X_i containing both x and y. Also, for every vertex x of C the set of subsets X_i containing x forms a subtree of T. The lemma follows since a family of subtrees of a tree satisfies the Helly property (see Definition 2.1.4).   □

Given a tree-decomposition D of a graph G of width k, we can triangulate G into a triangulated graph H with maximal clique number $k + 1$ as follows.

Definition 2.2.6 *Let* $D = (S, T)$ *be a tree-decomposition for* $G = (V, E)$. *Define* $H = \mathcal{H}(G, D)$ *as the graph with the same vertex set as G, with two vertices in H adjacent if and only if they appear in some common subset* $X \in S$. *We call H the* triangulation *of G implied by D.*

Lemma 2.2.3 *Let* $D = (S, T)$ *be a tree-decomposition for* $G = (V, E)$ *of width* k. *Then* $H = \mathcal{H}(G, D)$ *is a triangulation of* G *and has clique number* $k + 1$. *The tree-decomposition* D *is also a tree-decomposition for* H.

Proof. First we show that G is a subgraph of H. If (v, w) is an edge in G, then v and w appear in some common subset $X \in S$. By definition v and w are thus also adjacent in H. The fact that D is also a tree-decomposition for H follows immediately from Definition 2.2.2. By Lemma 2.2.2 the clique number of H is $k + 1$.

It remains to show that H is triangulated. For each vertex $x \in V$, consider the subtree of T consisting of those nodes i for which $x \in X_i$. Two vertices x and y are adjacent in H if and only if the corresponding subtrees intersect. Hence, H is the intersection graph of a family of subtrees of a tree, and by Lemma 2.1.5 H is triangulated.   □

An alternative way to prove Lemma 2.2.3 is to show that H has a perfect elimination scheme. This can be seen as follows. First, if T has only one node, H is a clique and thus H is triangulated. If T has more than one node, let i be a leaf and let j be the neighbor of i. Assume that for every vertex $x \in X_i$ there exists another subset X_k ($k \neq i$) that also contains x. Then, by definition, x must also be contained in X_j and thus $X_i \subset X_j$. Now we can make a new tree-decomposition D' for H by removing X_i from the set of subsets and by removing i from the tree. Otherwise, let x be a vertex in X_i which is contained only in X_i. Then clearly all neighbors of x are contained in X_i and thus x is simplicial. In this way we obtain a perfect elimination scheme for H, showing that H is triangulated.

Corollary 2.2.1 *If* G *has a tree-decomposition of width* k, *then the treewidth of* G *is at most* k.

The next lemma shows the equivalence between the two concepts.

Lemma 2.2.4 *The treewidth of a graph* G *equals the minimum width over all tree-decompositions of* G.

Proof. Let the treewidth of G be k. We show how to construct a tree-decomposition of width k for G. Since the treewidth of G is k we know there exists a k-tree H such that G can be triangulated into H. First notice that a tree-decomposition for H is also a tree-decomposition for G, since G is a subgraph of H with the same set of vertices. Hence it is sufficient to show there exists a tree-decomposition for H of width k.

Since H is triangulated, it is the intersection graph of a family of subtrees of a tree $T = (I, F)$. For each node i in this tree, define a subset X_i consisting of those vertices for which the corresponding subtree contains i. Let $S =$

$\{X_i \mid i \in I\}$. We claim that $D = (S, T)$ is a tree-decomposition of width
at most k. It is easy to check that D is indeed a tree-decomposition for
H. Furthermore, each subset corresponds with a clique in H and hence has
cardinality at most $k + 1$. This shows that the width of D is at most k.

The converse is stated in Corollary 2.2.1. $\square$

Lemma 2.2.5 *For any graph with* n *vertices and treewidth* k, *there exists
a tree-decomposition* $D = (S, T)$ *with* $|S| \leq n - k$ *such that every subset
$X \in S$ contains exactly* $k + 1$ *vertices.*

Proof. Let $G = (V, E)$ be a graph with n vertices and treewidth k. Let H be
a triangulation of G into a k-tree. We show there exists a tree-decomposition
as claimed for H. If $n = k + 1$, then we take a tree with one node and a
corresponding subset containing all nodes. Otherwise let x be a simplicial
vertex. By induction there exists a tree-decomposition for $H[V \setminus \{x\}]$ with
$n - k - 1$ nodes and such that every subset in the tree-decomposition has
$k + 1$ elements. Since $N(x)$ is a clique, there is a subset X_i in this tree-
decomposition such that $C \subseteq N(x)$. Take a new node i' and make this
adjacent to i. Let the corresponding subset be $X_{i'} = \{x\} \cup N(x)$. $\square$

For a related result see also Lemma 7.2.3 on page 81.

We now show some related results for the path-decomposition of a
graph. In the same manner in which triangulated graphs are related to
tree-decompositions, interval graphs are related to path-decompositions.

Definition 2.2.7 *A* k-path *is a* k-tree *which is an interval graph. A
partial* k-path *is a subgraph of a* k-path. *The* pathwidth *of a graph* G *is
the minimum value* k *for which* G *is a partial* k-path.

Analogous to Lemma 2.1.13 and Lemma 2.1.14 we can obtain the following
results.

Lemma 2.2.6 *An interval graph* G *with at least* $k + 1$ *vertices and with
$\omega(G) \leq k + 1$ can be triangulated into a* k-path.

Proof. Let $G = (V, E)$ be an interval graph with $n \geq k + 1$ vertices and
with $\omega(G) \leq k + 1$. Let $(X_1, \ldots, X_t)$ be an interval ordering of the maximal
cliques of G. We prove there exists a triangulation H of G into a k-path
such that there is an interval ordering $(Y_1, \ldots, Y_{n-k})$ of the maximal cliques
of H, with $X_1 \subseteq Y_1$. This can be seen as follows. If G has $k + 1$ vertices
then we can take for H a clique with all vertices and let $Y_1 = V$. Otherwise
G can not be a clique and there must exist a vertex $x \in X_1 \setminus X_2$. Then x
is a simplicial vertex in G and $G[V \setminus \{x\}]$ is an interval graph with at least
$k + 1$ vertices and with clique number at most $k + 1$. There are two cases
to consider.

First assume that $(X_1 \setminus \{x\}, X_2, \ldots, X_t)$ is an interval ordering of the maximal cliques of $G[V \setminus \{x\}]$. By induction there is a k-path H', which is a triangulation of $G[V \setminus \{x\}]$, and there is an interval ordering $(Y_1, \ldots, Y_{n-k-1})$ with $X_1 \setminus \{x\} \subseteq Y_1$. Notice there must be an element $z \in Y_1 \setminus X_1$ (otherwise X_1 contains more than $k+1$ vertices). Define $Y_0 = \{x\} \cup (Y_1 \setminus \{z\})$. Let H be the graph obtained from H' by making x adjacent to all vertices of $Y_0 \setminus \{x\}$. Then H is a k-tree, and $(Y_0, Y_1, \ldots, Y_{n-k-1})$ is an interval ordering of the maximal cliques of H (hence H is a k-path) with $X_1 \subseteq Y_0$.

Now assume $X_1 \setminus \{x\}$ is not a maximal clique in $G[V \setminus \{x\}]$. Then $X_1 \setminus \{x\} \subset X_2$. Clearly $(X_2, \ldots, X_t)$ is an interval ordering of the maximal cliques of $G[V \setminus \{x\}]$. Let H' be a triangulation of $G[V \setminus \{x\}]$ into a k-path, and let $(Y_2, \ldots, Y_{n-k-1})$ be an interval ordering of the maximal cliques of H with $X_2 \subseteq Y_2$. Let $z \in Y_2 \setminus X_1$ (this vertex must exist since $X_2 \setminus X_1 \neq \emptyset$). Let $Y_1 = \{x\} \cup (Y_2 \setminus \{z\})$. Let H be the graph obtained from H' by making x adjacent to all vertices of $Y_1 \setminus \{x\}$. Then H is a k-tree and $(Y_1, \ldots, Y_{n-k})$ is an interval ordering of the maximal cliques of H with $X_1 \subseteq Y_1$.   □

Lemma 2.2.7 *Every partial k-path with at least $k+1$ vertices can be triangulated into a k-path.*

Proof. Let $G = (V, E)$ be a partial k-path with at least $k+1$ vertices. There is a k-path H such that G is a subgraph of H. Notice that $H[V]$ is an interval graph with maximal clique number at most $k+1$, such that G is a subgraph of $H[V]$. By Lemma 2.2.6, $H[V]$ can be triangulated into a k-path.   □

Lemma 2.2.8 *The pathwidth of a graph G equals the minimum width over all path-decompositions of G.*

Proof. Assume the pathwidth of G is k. Hence G is the subgraph of an interval graph H which is simultaneously a k-tree. Since H is a k-tree all maximal cliques of H have number $k+1$. We show there exists a path-decomposition for H of width k. By Lemma 2.1.6 there is an interval ordering $(X_1, X_2, \ldots, X_\ell)$ of the maximal cliques of H. It is easy to check that $(X_1, X_2, \ldots, X_\ell)$ is a path-decomposition for H of width k.

Conversely let $D = (X_1, X_2, \ldots, X_\ell)$ be a path-decomposition for G of width k. Let H be the triangulation of G implied by D. Each maximal clique of H is contained in some subset X_i and each subset contains at most one maximal clique. Hence there exists an ordering of the maximal cliques of H such that for every vertex the maximal cliques containing it are consecutive. Hence H is an interval graph with clique number $k+1$. It follows that H can be triangulated into a k-path.   □

We show that the treewidth and pathwidth of a graph differ at most by a factor $\log n$. We need the following lemma. For a slightly more general result see also [143]. See also Lemma 2.1.4.

Lemma 2.2.9 *Every k-tree* $G = (V, E)$ *contains a clique* C *with* $k + 1$ *vertices such that every connected component of* $G[V \setminus C]$ *has at most* $\frac{1}{2}(n - k)$ *vertices.*

Proof. Consider the following algorithm. Start with any $k + 1$-clique S_0. Assume there is a connected component C in $G[V \setminus S_0]$ which has more than $\frac{1}{2}(n - k)$ vertices. Notice that the other components together have less than $\frac{1}{2}(n - k) - 1$ vertices. There exists a vertex x in C which has k neighbors in S_0. Let $y \in S_0 \setminus N(x)$. Define $S_1 = \{x\} \cup (N(x) \cap S_0)$. Notice that S_1 also has $k + 1$ vertices. The algorithm continues with S_1.

To show that this algorithm terminates, we prove that in each step of the algorithm the number of vertices in the largest component decreases. Notice that $G[V \setminus S_1]$ has two types of components. One type consists only of vertices of $C \setminus \{x\}$. If the largest component of $G[V \setminus S_1]$ is among these, the number of vertices has clearly decreased. The other type of components consists only of vertices of $\{y\} \cup V \setminus (C \cup S_0)$. By the remark above, the total number of vertices in this set is less than $\frac{1}{2}(n - k)$. As the largest component of $G[V \setminus S_0]$ has more than this number of vertices, this shows that the number of vertices in the largest component of $G[V \setminus S_1]$ is at least one less than the number of vertices in the largest component of $G[V \setminus S_0]$. This shows that the algorithm terminates. $\qquad\Box$

Corollary 2.2.2 *Let* G *be a graph with* n *vertices and treewidth* k. *There exists a set* S *with* $k + 1$ *vertices such that every connected component of* $G[V \setminus S]$ *has at most* $\frac{1}{2}(n - k)$ *vertices.*

Lemma 2.2.10 *Let* $G = (V, E)$ *be a graph with* n *vertices and treewidth* $k \geq 1$. *Then the pathwidth of* G *is at most* $(k + 1) \log n - 1$.

Proof. Lemma 2.2.9 shows there is a clique X with $k + 1$ vertices such that every connected component of $G[V \setminus X]$ has at most $\frac{1}{2}(n - k)$ vertices.

If $n = k + 1$ the upperbound on the pathwidth clearly holds. We proceed by induction. Let $n > k + 1$, and let X be a balanced separator as mentioned above. Let $C_1, \ldots, C_t$ be the connected components of $G[V \setminus X]$. By induction there are path-decompositions P_i for the induced subgraphs $G[C_i]$ with pathwidth $\leq (k + 1) \log |C_i| - 1$. Add X to every subset of every path-decomposition, and let P'_i $(i = 1, \ldots, t)$ be the new path-decompositions so obtained. Let P be the concatenation of the P'_i's, i.e., the path-decomposition obtained by putting all P'_i's after each other: $P = P'_1 + + \ldots + + P'_t$. Clearly, the width of P is at most $k + (k + 1) \log \frac{n-k}{2} \leq (k + 1) \log n - 1$. $\qquad\Box$

2.3   Perfect graphs

Perfect graphs were introduced by Claude Berge around 1960 (see [13]). A graph $G = (V, E)$ is called *perfect* if the following two conditions are both satisfied: First the *clique number* and the *chromatic number* must be equal for all induced subgraphs, (i.e. $\omega(G[A]) = \chi(G[A])$ for all $A \subseteq V$) and second, the *stability number* must equal the *clique cover number* for all induced subgraphs of G (i.e. $\alpha(G[A]) = k(G[A])$ for all $A \subseteq V$). Notice that it is quite a natural question to ask which graphs are perfect, since all graphs satisfy $\omega(G) \leq \chi(G)$ and $\alpha(G) \leq k(G)$. Notice also that the two conditions are dual in the sense that a graph satisfies the first condition if and only if its complement satisfies the second. The remarkable fact that a graph satisfies the first equality if and only if it satisfies the second equality, was conjectured by Berge [12] and proven by Lovász [121]. This has become known as the *perfect graph theorem*. Lovász also proved that the two conditions are equivalent to a third condition, namely: $\omega(G[A])\alpha(G[A]) \geq |A|$ for all $A \subseteq V$.

A graph is called *minimal imperfect* (or p-*critical*) if it is not perfect but every proper induced subgraph of it is. The *strong perfect graph conjecture* made by Berge (see [14]) states that the only minimal imperfect graphs are the chordless odd cycles of length at least five and their complements. This is equivalent to saying that a graph G is perfect if and only if in G and in $\overline{G}$ every chordless odd cycle of length at least five has a chord. The chordless odd cycles of length at least five and their complements are often referred to as the odd holes and the odd antiholes, respectively. Until now, the strong perfect graph conjecture is unsettled.

It is interesting to notice that for every constant k there is a polynomial time algorithm to test whether a given partial k-tree satisfies the strong perfect graph conjecture. This can be seen as follows. Let G be a partial k-tree. We may assume that a tree-decomposition of G of width bounded by some constant is given. Using standard techniques it can be tested in linear time whether or not the graph is perfect (for example, it can be stated in monadic second order logic [49] or, it can be tested using dynamic programming). Also, testing whether the graph has an odd hole can be done in linear time using one of these techniques. The only thing left to test is whether the graph has an odd antihole. Now notice that an antihole $\overline{C_{2t+1}}$ has a clique of number at least t. Since a partial k-tree can only have cliques of size at most $k+1$ it follows that we can restrict the search for odd antiholes to those with at most $2k + 3$ vertices. It follows that we can perform this test also in linear time. In chapter 14 we show a linear time algorithm to obtain an optimal tree-decomposition for the graph. This proves that for each k there is a linear time algorithm to test whether a given partial k-tree satisfies the strong perfect graph conjecture. Using results of [48] we can even obtain a much stronger result; for each k there is an algorithm to decide whether or not for *every* partial k-tree the strong perfect graph conjecture holds. We do not claim that this is a very practical algorithm. This is even more so since for partial 3-trees the strong perfect graph conjecture is already known to hold. This can be seen as follows. Notice that every graph with at most three vertices is either bipartite or a clique. Consider a partial k-tree for $k \leq 3$ which is minimal imperfect. Then there is a vertex u with at most three neighbors. Hence the neighborhood of this vertex induces a bipartite graph or a clique (which is multipartite). In [87] it is shown that every minimal imperfect graph which has a vertex of which the neighborhood induces a bipartite or multipartite graph, must be an odd hole or an odd antihole.

Since the discovery of perfect graphs in 1960, much research has been devoted to special classes of perfect graphs. Among the most well-known classes of perfect graphs are the comparability graphs and the triangulated graphs. The class of triangulated graphs contains graph classes such as interval graphs, split graphs, k-trees, and indifference graphs. The class of comparability graphs contains complements of interval graphs, permutation graphs, threshold graphs and P_4-free graphs (or cographs). Much work has been done to characterize these graph classes and to find relationships between them. Interest has only increased since Lovász settled the perfect graph conjecture. From an algorithmic point of view, perfect graphs have become of great interest since Grötschel, Lovász and Schrijver discovered polynomial time algorithms for NP-hard problems like CLIQUE, STABLE SET and CHROMATIC NUMBER for perfect graphs [15]. Special classes of perfect graphs have proven their importance by the large number of applications (see for example [76, 15, 141] for applications in general and [46] for an overview of applications of interval graphs).

Chapter 3

Testing superperfection of k-trees

Much work has been done in recognizing classes of perfect graphs in polynomial time [35, 72, 76, 135, 155]. An exception seems to be the class of superperfect graphs.

The results presented in this chapter can be summarized as follows. First we give in section 3.2 a complete characterization, by means of forbidden induced subgraphs, of 2-trees that are superperfect. We also characterize those 2-trees that are comparability graphs and those that are permutation graphs. Secondly, in section 3.3 we give *for each constant* k an $O(1)$ time algorithm which produces a complete characterization of superperfect k-trees, by means of forbidden configurations. With the aid of this characterization we find, for each constant k, a linear-time algorithm to test superperfection of k-trees.

Until now we have not been able to find a polynomial algorithm to test superperfection on partial k-trees (for general k). Since by definition a graph is superperfect if for each assignment of non-negative weights to the vertices the interval chromatic number is equal to the maximum weight clique, the following observation is of interest. Determining the interval chromatic number of a weighted interval graph with weights one and two is NP-hard. When restricted to weighted partial k-trees, for some constant k, it can be seen that the interval chromatic number can be determined in linear time.

The class of superperfect graphs contains the class of comparability graphs (hence also the class of bipartite graphs), but these classes are not equal. This has been pointed out by Golumbic [76] who showed the existence of an infinite class S of superperfect graphs which are not comparability graphs. However all graphs in S are neither triangulated nor co-triangulated, and therefore Golumbic [76] raises the question whether for triangulated graphs the classes of superperfect and comparability graphs coincide. For split graphs this equivalence has been shown. Our results show it is not the case for triangulated graphs in general.

3.1   Preliminaries

We start with some definitions and easy lemmas. Most definitions and results
in this section are taken from [76]. For further information on superperfect
graphs the reader is referred to this book.

Definition 3.1.1 *An undirected graph* $G = (V, E)$ *is called a comparability
graph, or a transitively orientable graph, if there exists an orientation
of the edges such that the resulting oriented graph* (V, F) *satisfies the
following conditions.*

$$F \cap F^{-1} = \emptyset \ and \ F + F^{-1} = E \ and \ F^2 \subseteq F$$

where $F^2 = \{(a, c) \mid \exists_{b \in V} (a, b) \in F \wedge (b, c) \in F\}$. *An orientation* F *of the
edges satisfying the conditions above is called a* transitive *orientation.*

So, if F is a transitive orientation, then $(a, b) \in F$ and $(b, c) \in F$ imply
$\{a, c\}$ is an edge with orientation $(a, c) \in F$. There exists a somewhat weaker
equivalent condition (which we do not use): a graph is a comparability graph
if and only if it admits an orientation of its edges that represents a pseudo-
order relation (see [15], page 76).

If a graph G is a comparability graph, then this also holds for every
induced subgraph of G. In [76] it is shown that comparability graphs are
perfect, and can be recognized in polynomial time (see also [155]).

A *weighted* graph is a pair (G, w), where G is a graph and w a weight
function which associates to every vertex x a non-negative weight $w(x)$. For
a subset S of the vertices we define the weight of S, denoted by $w(S)$, as
the sum of the weights of the vertices in S.

Definition 3.1.2 *An* interval coloring *of a weighted graph* (G, w) *maps
each vertex* x *to an open interval* I_x *on the real line, of width* $w(x)$, *such
that adjacent vertices are mapped to disjoint intervals. The* total width
of an interval coloring is defined to be $|\bigcup_x I_x|$. *The* interval chromatic
number $\chi(G, w)$ *is the least total width needed to color the vertices with
intervals.*

Determining whether $\chi(G, w) \leq r$ is an NP-complete problem, even if w is
restricted to values one and two and G is an interval graph. (This has been
shown by L. Stockmeyer as reported in [76].) In this paper we shall only use
the following alternative characterization of the interval chromatic number
(see [76]).

Theorem 3.1.1 *If* (G, w) *is a weighted undirected graph, then*

$$\chi(G, w) = \min_F \left(\max_\mu w(\mu) \right)$$

where F *is an acyclic orientation of* G *and* μ *is a path in* F.

If w is a weight function and F is an acyclic orientation, then we say that F is a *superperfect orientation* with respect to w if the weight of the heaviest path in F does not exceed the weight of the heaviest clique.

Definition 3.1.3 *The* clique number $\Omega(G, w)$ *of a weighted graph* (G, w) *is defined as the maximum weight of a clique in* G.

In this chapter we use the capital Ω to denote the (weighted) clique number rather then ω, to avoid confusion with the weight function w. It is easy to see that $\Omega(G, w) \le \chi(G, w)$ holds for all weighted graphs, since for any acyclic orientation and for every clique there exists a path in the orientation which contains all vertices of the clique.

Definition 3.1.4 *A graph* G *is called* superperfect *if for every non-negative weight function* w, $\Omega(G, w) = \chi(G, w)$.

Notice that each induced subgraph of a superperfect graph is itself superperfect, and also that every superperfect graph is perfect. If G is a comparability graph, then there exists an orientation such that every path is contained in a clique. This proves the following theorem (see also [76]).

Theorem 3.1.2 *Every comparability graph is superperfect.*

The converse of this theorem is not true. In [76] an infinite class of superperfect graphs is given that are not comparability graphs. However, none of these graphs is triangulated. In [76] (page 214) the question is raised whether the converse of the theorem holds for triangulated graphs; is it true or false that, for *triangulated* graphs, G is a comparability graphs if and only if G is superperfect? In the next section we answer this question in the negative, and we give a complete characterization of superperfect 2-trees.

3.2 2-trees and superperfection

In this section we give a characterization of the 2-trees that are superperfect by means of forbidden subgraphs. In 1967 Gallai [68] published a list of all minimal forbidden subgraphs of the comparability graphs (see also [15] page 78). Extracting from this list the triangulated graphs which are subgraphs of 2-trees (or: have treewidth at most two), we find a characterization of the 2-trees which are comparability graphs. We find two types of forbidden induced subgraphs, which we call the 3-*sun* and the *odd wing*. They are illustrated in figure 3.1. Notice that a 3-sun and a wing are 2-trees, and that a wing has at least seven vertices. We call a wing odd (even) if the total number of vertices is odd (even). The following lemma is easy to check.

Lemma 3.2.1 *A wing is a comparability graph if and only if it is even.*

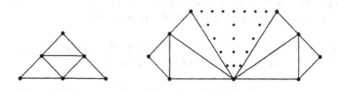

Figure 3.1: 3-sun (left) and wing (right)

We thus find the following characterization of 2-trees that are comparability graphs.

Theorem 3.2.1 *A 2-tree is a comparability graph if and only if it does not contain a 3-sun or an odd wing.*

It is interesting to notice that we can get a characterization of the 2-trees that are permutation graphs. Recall that a graph is an interval graph if and only if it is triangulated and a cocomparability graph [72]. Also, a graph is a permutation graph if and only if the graph and its complement are comparability graphs [135]. It follows that a 2-tree is a permutation graph if and only if it is an interval graph without an induced odd wing.

The next theorem shows that the smallest odd wing, with seven vertices, (which is not a comparability graph) is superperfect. As we shall see later, this is in fact the only odd wing that is superperfect. Notice that in [76] (page 212, figure 9.9) this graph is mistakenly placed in the position of a non-superperfect graph. See also [125] and [62]; the result of [125] is wrong: A wing is an interval graph.

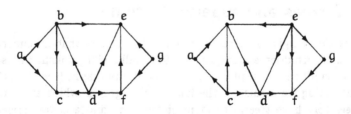

Figure 3.2: two orientations of the wing with seven vertices

Theorem 3.2.2 *The odd wing with seven vertices is superperfect and hence there exists a triangulated graph which is superperfect but not a comparability graph.*

Proof. Label the vertices of the graph as in figure 3.2. We consider two orientations of this wing as illustrated in figure 3.2 and we show that for every weight function one of these orientations is superperfect. Notice that both orientations are such that there is exactly one path not contained in a triangle. In the first orientation this is the path $\{a, b, e\}$ and in the second orientation the path $\{c, d, f\}$. Consider a non-negative weight function w of the vertices. Suppose the orientation of the first type is *not* superperfect with respect to w. Then the path $\{a, b, e\}$ must be heavier then every triangle. Since $\{a, b, c\}$ is a triangle, this implies that $w(e) > w(c)$. But then $w(\{c, d, f\}) < w(\{e, d, f\})$, and since $\{e, d, f\}$ is a triangle, the second orientation is superperfect with respect to w.   □

In the last part of this section we give a complete characterization of the superperfect 2-trees. In figure 3.3 we give a list of forbidden induced subgraphs. Notice that the fourth subgraph starts an infinite series. The following lemma can be easily checked.

Lemma 3.2.2 *The graphs shown in figure 3.3 are not superperfect. For each of the graphs the weight function that is shown is such that for any acyclic orientation, there exists a path which is heavier than the heaviest clique.*

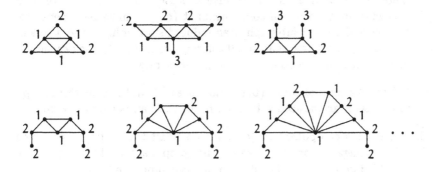

Figure 3.3: critical non-superperfect graphs

Theorem 3.2.3 *A 2-tree is superperfect if and only if it does not contain an induced subgraph isomorphic to one of the graphs shown in figure 3.3.*

Proof. Assume the 2-tree G has no induced subgraph from this list. Then the graph cannot have an induced odd wing with nine or more vertices. We may assume the graph is not a comparability graph, hence it contains an odd wing with seven vertices. Consider the labeled wing of figure 3.2. Let H be the subgraph obtained from this wing by removing the vertices a and g. Let (x, y) be an edge of H. We say that C is a component at (x, y) if C is a maximal connected subgraph of $G[V \setminus \{x, y\}]$ *containing no vertices of* H. If C is a component at (x, y), we say that x is a degenerate vertex of this component if x is adjacent to all vertices of C. Notice that there are only four edges of H at which there can be components, namely (b, c), (b, d), (d, e) and (e, f), otherwise the 3-sun would be an induced subgraph. The following remarks restrict the possible components.

1. It can be checked that if C is a component at (x, y), then either x or y must be degenerate.

2. For components at (b, d), b must be degenerate and for components at (d, e), e must be degenerate, otherwise the second graph in the list is an induced subgraph.

3. Consider components at (b, c) with at least two vertices. Either b or c is degenerate for *all* these components, otherwise the second graph from list 3.3 is an induced subgraph.

4. If there is a component at (b, c) with at least two vertices for which c is degenerate, then for all components at (e, f), e must be degenerate, otherwise the fourth graph in the list is an induced subgraph. If there is a component at (b, c) with at least two vertices for which b is degenerate then for all components at (e, f), f must be degenerate, otherwise the third subgraph is an induced subgraph. Without loss of generality, we assume that for all components at (b, c), c is degenerate and for all components at (e, f), e is degenerate.

5. If there is a component at (b, d) and one at (d, e), then the third graph in the list is an induced subgraph, hence this can not be the case.

6. If there is a component at (e, f) with at least two vertices (for which e is degenerate) then there can be no component at (b, d), otherwise the third graph from the list is an induced subgraph.

7. If there is a component at (b, c) with at least two vertices, then all components at (b, d) can have only one vertex, otherwise the second graph of the list is an induced subgraph.

8. Suppose there is a component at (b, d). Then all components at (b, c) can have at most two vertices. Furthermore, if there is a component at (b, c) with two vertices, then it is the only component at (b, c). Otherwise, the third subgraph of the list is an induced subgraph.

It follows that only one of two cases can occur.

- There is a component at (b, d). Then there is no component at (d, e). All components at (b, d) and at (e, f) have one vertex. Either all components at (b, c) have one vertex or there is only one component at (b, c), in which case it can have at most two vertices and c is degenerate.

- There are only components at (b, c), (d, e) and (e, f). For all components at (b, c), c is degenerate. For all components at (d, e) and at (e, f), e is degenerate.

It is easily checked (e.g. by methods described in the next section) that both types are superperfect. □

Notice that the list of forbidden induced subgraphs is infinite. In the next section we show that for each k there is a finite characterization of the class of superperfect k-trees, by means of forbidden configurations. Furthermore we give for each k a $O(1)$ time algorithm to find this characterization. As a consequence we find a linear time algorithm to check whether a k-tree is superperfect.

3.3 k-trees and superperfection

In this section, let k be some constant, and let G be a k-tree. We start by showing that we can restrict the set of orientations if we want to test whether G is superperfect. In this chapter a *coloring* of a graph $G(V, E)$ with $k + 1$ colors is a function $C : V \rightarrow \{1, \ldots, k+1\}$, such that $C(x) \neq C(y)$ whenever x and y are adjacent. We only use colorings with $k + 1$ colors; we do not always mention the number of colors. A coloring of a k-tree is unique up to a permutation of the colors, as is easily seen by induction.

Lemma 3.3.1 *If C and C' are two colorings of a k-tree $G = (V, E)$ then there exists a permutation π of the colors $\{1, \ldots, k+1\}$ such that for every vertex x, $C(x) = \pi(C'(x))$.*

By this fact, the following set of orientations is uniquely defined for each k-tree.

Definition 3.3.1 *Let G be a graph and let C be a coloring of G with $k+1$ colors. For each permutation π of the colors we define an orientation F_π as follows. Direct the edge (x, y) from x to y if $\pi(C(x)) < \pi(C(y))$. Let $\mathcal{F}_c(G)$ be the set of orientations obtained in this way.*

The following lemma is an immediate consequence of Definition 3.3.1.

Lemma 3.3.2 *If* G *is a* k-tree *then:*

1. $|\mathcal{F}_c| = (k+1)!$.

2. *Each* $F \in \mathcal{F}_c$ *is acyclic.*

3. *If* $F \in \mathcal{F}_c$ *then each path* μ *in* F *has at most* $k+1$ *vertices.*

Definition 3.3.2 *Let* G *be a* k-tree. *Let* $\mathcal{F}^*(G)$ *be the set of those acyclic orientations of* G *in which every path contains at most* $k+1$ *vertices.*

Notice that $\mathcal{F}_c \subseteq \mathcal{F}^*$.

Lemma 3.3.3 $\mathcal{F}_c = \mathcal{F}^*$.

Proof. We prove that $|\mathcal{F}^*| = (k+1)!$ which implies the result. Let $F \in \mathcal{F}^*$. Let S be a k-clique in G, and let x and y be two vertices which are adjacent to all vertices of S. Since S is a clique and F is acyclic, there is a unique ordering of the vertices of S, say $s_1, s_2, \ldots, s_k$, such that there is an arc from s_i to s_j ($s_i \to s_j$) if and only if $i > j$. Since x is adjacent to all vertices of S, there exists an index $0 \le t_x \le k$ such that $x \to s_i$ for all $1 \le i \le t_x$ and $s_j \to x$ for all $t_x < j \le k$. The same holds for y with index t_y. Consider the case $t_x < t_y$. Then F has a path of length $k+2$:

$$\left(s_k, s_{k-1}, \ldots, s_{t_y+1}, y, s_{t_y}, \ldots, s_{t_x+1}, x, s_{t_x}, \ldots, s_1 \right)$$

Since $F \in \mathcal{F}^*$, we find that $t_x = t_y$. Now consider the inductive construction of G as a k-tree. Start with an acyclic orientation of a $(k+1)$-clique. This can be done in $(k+1)!$ manners. If we add a new vertex v and make it adjacent to a k-clique, by the argument above, the orientations of the edges incident with v are determined. Hence $|\mathcal{F}^*| = (k+1)!$.   □

Definition 3.3.3 *Let* F *be an acyclic orientation. A path* μ *in* F *is contained in a path* μ', *if all vertices of* μ *are also vertices of* μ'.

Lemma 3.3.4 *Let* $F \in \mathcal{F}_c$. *Then any path* μ *in* F *is contained in a path* μ' *with* $k+1$ *vertices.*

Proof. Let C be a coloring and let $F = F_\pi$ for some permutation π. The colors in the path μ must appear in the same order as in the permutation. Assume there is a gap between adjacent colors c_1 and c_2 in the path (i.e., there is a color in the permutation between c_1 and c_2). Since the edge of the path with colors c_1 and c_2 is contained in a $(k+1)$-clique, the missing colors can be put between c_1 and c_2. Thus we can make a longer path μ' containing μ.   □

Theorem 3.3.1 *Let G be a k-tree. Then G is superperfect if and only if for all weight functions w*

$$\min_{F \in \mathcal{F}_c} \max_{\mu} w(\mu) = \Omega(G, w)$$

Proof. Assume G is superperfect. Let w be a non-negative weight function. There is an orientation F such that $\max_{\mu} w(\mu) = \Omega(G, w)$. If every path in F has at most $k+1$ vertices, we are done. Assume F has a path with more than $k+1$ vertices. Now increase all weights with some constant $L > \Omega(G, w)$. Let w' be this new weight function. Notice that $\Omega(G, w') = \Omega(G, w) + (k+1)L$. Since G is superperfect, there must be an orientation F′ for this new weight function w'. Suppose F′ also has a path μ with more than $k+1$ vertices. Then (with $|\mu|$ the number of vertices of μ):

$$
\begin{aligned}
\Omega(G, w') = \Omega(G, w) + (k+1)L &\geq w'(\mu) \\
&= w(\mu) + |\mu|L \\
&\geq w(\mu) + (k+2)L \\
&\geq (k+2)L
\end{aligned}
$$

Since $L > \Omega(G, w)$, this is a contradiction. We may conclude that $F' \in \mathcal{F}^* = \mathcal{F}_c$. We show that F′ is also a good orientation for the weight function w. Let ν be a path in F′. By Lemma 3.3.4, ν is contained in a path ν^* with $k+1$ vertices. Hence

$$w(\nu) \leq w(\nu^*) = w'(\nu^*) - (k+1)L \leq \Omega(G, w') - (k+1)L = \Omega(G, w)$$

The converse is trivial. □

Definition 3.3.4 *Let G be a triangulated graph and let C be a coloring of G with $k+1$ colors. For each permutation π of the colors, let $\mathcal{P}(\pi)$ be the set of paths in F_π, which are not contained in a clique and which have $k+1$ vertices. If Q is a set of paths in G, we say that Q is a cover if, for every permutation π, there is a path $\mu \in Q$ which can be oriented such that it is in $\mathcal{P}(\pi)$. A cover is called minimal if it contains $\frac{1}{2}(k+1)!$ paths.*

Lemma 3.3.5 *Let G be a k-tree. If for some permutation π, $\mathcal{P}(\pi) = \emptyset$, then G is a comparability graph (hence superperfect).*

Proof. Suppose $\mathcal{P}(\pi) = \emptyset$. Consider the orientation F_π. If there is a path in F_π which is not contained in a clique then, by Lemma 3.3.4, $\mathcal{P}(\pi)$ cannot be empty. Hence, every path in F_π is contained in a clique. Since F_π is acyclic, the lemma follows. □

Definition 3.3.5 *Let* G *be a* k-*tree and let* C *be a coloring of* G. *Let* S *be a maximal clique of* G, *and let* Q *be a minimal cover. Define* $LP(G, S, Q)$ *as the following set of inequalities:*

1. *For each vertex* x: $w(x) \geq 0$.

2. *For each maximal clique* $S' \neq S$: $w(S') \leq w(S)$.

3. *For each path* μ *of* Q: $w(\mu) > w(S)$.

We call the second type of inequalities, the clique inequalities. *The inequalities of the third type are called the* path inequalities.

Lemma 3.3.6 *There are* $n - k - 1$ *clique inequalities and* $\frac{1}{2}(k+1)!$ *path inequalities.*

Proof. Notice that a k-tree has $n - k$ cliques with $k + 1$ vertices.   □

Theorem 3.3.2 *Let* G *be a* k-*tree with a coloring* C. G *is not superperfect if and only if there is a maximal clique* S *and a minimal cover* Q *such that* $LP(G, S, Q)$ *has a solution.*

Proof. Suppose $LP(G, S, Q)$ has a solution. Take this solution as a weight function. Then, clearly, for any orientation F_π there is a path (in Q and in $P(\pi)$) which is heavier than the heaviest clique S. By Theorem 3.3.1 G is not superperfect. On the other hand, if G is not superperfect, there exists a weight function w such that for every orientation F_π there is a path which is heavier than the heaviest clique (hence it can not be contained in a clique). By Lemma 3.3.4 we may assume this path has $k + 1$ vertices, hence it is in $P(\pi)$. Take S to be the heaviest clique and let Q be a minimal cover for these paths.   □

Notice that we could use Theorem 3.3.2 for a polynomial time algorithm to test superperfection on k-trees. There are at most n^{k+1} different paths of length k, hence the number of minimal covers is at most $(n^{k+1})^{\frac{1}{2}(k+1)!}$. Since the number of maximal cliques of G is at most $n - k$, we only have to check for a polynomial number of sets of inequalities if it has a solution. This checking can be done in polynomial time, e.g. by the ellipsoid method. We now show that there also exists a linear time algorithm.

Consider a set of inequalities $LP(G, S, Q)$ which has a solution. Notice that if some vertex y does *not* appear in the path inequalities then we can set the weight $w(y) = 0$. This new weight function is also a solution. Hence we can transform the set of inequalities as follows:

Definition 3.3.6 *Let* G *be a* k*-tree and let* C *be a coloring of* G*. Let* S *be a maximal clique, and let* $\mathcal{Q}$ *be a cover. Let* H *be the subgraph of* G *induced by the vertices of* S *and of all paths in* $\mathcal{Q}$*. Define* LP'(H, S, $\mathcal{Q}$) *as the following set of inequalities:*

1. *For each vertex* x *of* H*:* $w(x) \geq 0$*.*

2. *For each maximal clique* $S' \neq S$ *of* H*:* $w(S') \leq w(S)$*.*

3. *For each path* μ *in* $\mathcal{Q}$*:* $w(\mu) > w(S)$*.*

The following lemma follows directly from Definitions 3.3.5 and 3.3.6.

Lemma 3.3.7 LP(G, S, $\mathcal{Q}$) *has a solution if and only if* LP'(H, S, $\mathcal{Q}$) *has a solution.*

Lemma 3.3.8 *The number of inequalities of* LP'(H, S, $\mathcal{Q}$) *is bounded by a constant.*

Proof. There are $\frac{1}{2}(k+1)!$ paths in $\mathcal{Q}$, each involving $k+1$ variables. The clique S has also $k+1$ vertices, hence it follows that the subgraph H, has at most $(\frac{1}{2}(k+1)! + 1)(k+1)$ vertices. Since H is triangulated, the number of maximal cliques in H is bounded by the number of vertices. Hence the number of clique inequalities is bounded by $(\frac{1}{2}(k+1)! + 1)(k+1)$.   □

We now have the following algorithm to test superperfection of k-trees.

Algorithm to test superperfection of G

Step 1 Generate a list of all $(k+1)$-colored triangulated graphs H, with at most $(1 + \frac{1}{2}(k+1)!)(k+1)$ vertices, for which there exists:

  1. A maximal clique S with $k+1$ vertices

  2. A set $\mathcal{Q}$ of $\frac{1}{2}(k+1)!$ paths, which is a minimal cover,

  such that LP'(H, S, $\mathcal{Q}$) has a solution.

Step 2 Make a coloring of G (with $k+1$ colors).

Step 3 Check whether a graph H from the list is an induced subgraph of G (preserving colors). If G has a subgraph from the list then G is not superperfect, otherwise it is.

Notice that generating the list takes $O(1)$ time (if k is a constant). Since the subgraphs have constant size, we can check whether such a subgraph is an induced subgraph of G in linear time, using standard techniques for (partial) k-trees (see [5]).

Theorem 3.3.3 *The algorithm correctly determines whether* G *is super-perfect, and does so in linear time.*

Proof. Assume G is not superperfect. Let C be a coloring of G. By Theorem 3.3.2, $LP(G, S, Q)$ has a solution for some maximal clique S and some minimal cover Q. Take H the colored subgraph induced by vertices of S and of paths in Q. By Lemma 3.3.7, $LP'(H, S, Q)$ has a solution, so the subgraph H is in the list. Conversely, suppose the colored graph G has a colored induced subgraph H from the list. Then H has a clique S with $k + 1$ vertices and a minimal cover Q such that $LP'(H, S, Q)$ has a solution. Since H is an induced subgraph of G preserving colors, the clique S is also a maximal clique in G and the cover Q is also a cover for G. Since $LP(G, S, Q)$ has a solution, G can not be superperfect.   □

Notice that the list only has to contain those subgraphs H, of which every vertex is either in the maximal clique S or on some path in Q (with S and Q as defined in the algorithm). For reasons of simplicity, we left this detail out of the algorithm.

Chapter 4

Triangulating 3-colored graphs

In this chapter we consider the problem of determining whether a given colored graph can be triangulated such that no edges between vertices of the same color are added. This problem originated from the Perfect Phylogeny problem from molecular biology, and is strongly related with the problem of recognizing partial k-trees. In this chapter we give a simple linear time algorithm that solves the problem when there are three colors. We do this by using the structural characterization of the class of partial 2-trees.

Consider a graph of which the vertices are colored such that no two adjacent vertices have the same color. In this chapter we consider the problem to determine whether we can triangulate the graph such that in the triangulated graph no two adjacent vertices have the same color. The problem of triangulating a colored graph such that no two adjacent vertices have the same color is polynomially equivalent to the *Perfect Phylogeny problem*, see e.g. [92]. This problem, which is concerned with the inference of evolutionary history from DNA sequences, is of importance to molecular biologists. Recently, this problem was proven to be NP-complete [23]. In [129] it was shown by Morris, Warnow and Wimer that the problem is solvable in $O(n^{k+1})$ time for k-colored graphs. Another special case was solved in [92] by Kannan and Warnow. In this chapter we consider the problem, for the case in which there are at most three colors. In [93] Kannan and Warnow solved this problem by an $O(n\alpha(n))$ time algorithm. Recently, Kannan and Warnow improved on their algorithm, and found a variant that uses linear time but more than linear space. In this chapter we give another, in our opinion much simpler, linear time algorithm which uses only linear space. This work was done more or less simultaneously with, and independent from this recent result of Kannan and Warnow. Recently, it was reported to me by Nishizeki that they also found a linear time algorithm using only linear space [132] (in Japanese).

4.1   A characterization of partial 2-trees

In this section we first address the problem of characterizing partial 2-trees. A graph is a partial 2-tree if and only if all its biconnected components are partial 2-trees, so when characterizing partial 2-trees we can restrict ourselves to biconnected partial 2-trees.

Lemma 4.1.1 *Let* G *be a biconnected partial 2-tree. Let* $S = \{x, y\}$ *be a vertex separator such that* $G[V \setminus S]$ *has at least three connected components. Then in any triangulation of* G *into a a 2-tree* (x, y) *is an edge.*

Proof. Let x and y be vertices such that $G[V \setminus \{x, y\}]$ has at least three connected components. Since the graph is biconnected, both x and y have at least one neighbor in every connected component and so there are paths between x and y with internal vertices in every connected component. Assume that G can be triangulated into a 2-tree H such that (x, y) is not an edge in H. In a 2-tree every minimal vertex separator is an edge (Lemma 2.1.11). Every minimal vertex separator for x and y contains at least one vertex of every connected component of $G[V \setminus \{x, y\}]$ (otherwise there would be a path between x and y). So the minimal separator contains at least three vertices, which is a contradiction.   □

Definition 4.1.1 *The* cell-completion *of* G *is the graph* $\overline{G}$, *obtained from* G *by adding an edge between every pair* $\{x, y\}$ *for which* $G[V \setminus \{x, y\}]$ *has at least three connected components. A* cell *of* G *is a set of vertices which form a chordless cycle in the cell-completion.*

Note that by Lemma 4.1.1 the cell-completion $\overline{G}$ of a partial 2-tree G is a subgraph of every triangulation of G into a 2-tree.

Definition 4.1.2 *A* tree of cycles *is a member of the class of graphs* C_T *defined recursively as follows:*

1. *Every chordless cycle is an element of* C_T.

2. *Given an element* T *of* C_T *and a chordless cycle* C, *the graph obtained by identifying an edge and its two end-vertices of* C *with an edge and its two end-vertices of* T *is also in* C_T.

Call a tree of cycles T, a tree of k-cycles if all chordless cycles in T have length k. Then for example a 2-tree is a tree of 3-cycles (triangles). We show in the next theorem that a biconnected graph is a partial 2-tree if and only if its cell-completion is a tree of cycles. The following lemma will be useful.

Lemma 4.1.2 *Let G be a biconnected partial 2-tree and let G be triangulated into a 2-tree H. For any edge $e = (x, y)$ in H, the number of
common neighbors of x and y in H is at least three if and only if the
number of components in $G[V \setminus \{x, y\}]$ is at least three.*

Proof. Let $e = (x, y)$ be an edge in H such that x and y have at least
three common neighbors in H. Two of these common neighbors cannot be
adjacent in H otherwise there would be a 4-clique. Also, two of these common neighbors cannot be in the same connected component of $H[V \setminus \{x, y\}]$,
since e must be a minimal vertex separator for them. Hence the number
of components in $G[V \setminus \{x, y\}]$ is at least three. To prove the converse, let
the number of components of $G[V \setminus \{x, y\}]$ be at least three. Let C be a
component of $G[V \setminus \{x, y\}]$. G is biconnected hence there is a path between
x and y with internal vertices in C. Since (x, y) is an edge in H this implies
that, in H, x and y must have a common neighbor in C. This shows that
the number of common neighbors of x and y in H is at least three.   □

Throughout this chapter, the phrase 'triangulating a chordless cycle with m
vertices' will mean the addition of $m - 3$ edges (i.e., the *minimum* number)
to the cycle such that the new graph that is obtained is triangulated.

Theorem 4.1.1 *A biconnected graph G is a partial 2-tree if and only if
its cell-completion $\overline{G}$ is a tree of cycles. By triangulating each chordless
cycle of $\overline{G}$ in all possible ways we obtain all possible triangulations of
G into a 2-tree.*

Proof. The 'only if'-part can be seen as follows. We must show that $\overline{G}$ is a
tree of cycles. Consider a triangulation of $\overline{G}$ into a 2-tree H. Consider the
edges in H that are not in $\overline{G}$. Any such edge of H must be incident with
at least two triangles since $\overline{G}$ is biconnected. If there are *three* triangles
incident with an edge of H, the edge is also present in $\overline{G}$ (Lemma 4.1.2).
Since H is a tree of triangles and the only edges that are not in $\overline{G}$ are edges
that are incident with two triangles, it follows that $\overline{G}$ must be a tree of
cycles.

Conversely, let $\overline{G}$ be a tree of cycles. Then by triangulating each cycle
into a 2-tree, the resulting graph is a tree of triangles, and hence a 2-tree.

This proves the theorem.   □

4.2   c-**Triangulating** 3-**colored graphs**

Definition 4.2.1 *Let* t *be an integer. A* t*-colored graph is a graph* $G = (V, E)$ *together with a vertex coloring which is a mapping* $f : V \to S$ *such that*

1. *each vertex is colored with one of the colors such that no two adjacent vertices have the same color (i.e.,* $f(x) \neq f(y)$ *whenever* $(x, y) \in E$ *),*

2. $|S| = t$ *and each color is used at least once (i.e.,* f *is surjective).*

The problem is to triangulate G (if possible) without introducing edges between vertices which have the same color. If such a triangulation exists we call it a c-*triangulation* (c stands for colored).

Note that a graph can be c-triangulated if and only if all the biconnected components can be c-triangulated, and a c-triangulation of a graph G can be obtained by c-triangulating all biconnected components. Hence, in the remainder we assume that G is a biconnected graph. We can also observe that the clique number in a c-triangulation can be at most the number of different colors, since otherwise there would be an edge between vertices of the same color. The following lemma states an even stronger result.

Lemma 4.2.1 *Let* $G = (V, E)$ *be a* $(k + 1)$*-colored graph. Then* G *can be* c*-triangulated if and only if* G *can be* c*-triangulated into a* k*-tree.*

Proof. Let G be a $(k + 1)$-colored graph and assume that G can be c-triangulated. We show that G can also be c-triangulated into a k-tree. The converse is trivial: if G can be c-triangulated into a k-tree then, a fortiori, G can be c-triangulated.

We show, by induction on the number of vertices, that every $(k + 1)$-colored graph which is triangulated can be c-triangulated into a k-tree. This will prove the lemma since G can be c-triangulated into a triangulated graph H (which is $(k + 1)$-colored) and, by the statement above, H can be c-triangulated into a k-tree.

Notice that, since H is k + 1-colored, $n \geq k + 1$. If $n = k + 1$, then the lemma is obviously true. Assume $n > k + 1$. Since H is triangulated, it has a perfect elimination scheme $\sigma = [v_1, \ldots, v_n]$. Consider the graph obtained from H by removing the simplicial vertex v_1. Let C be the set of neighbors of v_1 in H. Hence C is a clique and $|C| \leq k$. By induction, $H[V \setminus \{v_1\}]$ can be triangulated into a k-tree H'. Since H' is a k-tree and C is a clique in H', C is contained in a $k + 1$-clique C'. Since C' is a clique with $k + 1$ vertices it contains exactly one vertex x with the same color as v_1. Clearly x cannot be contained in C. By making v_1 adjacent to every vertex of $C' \setminus \{x\}$ we obtain a c-triangulation of H into a k-tree.   □

In particular, it follows that a $(k + 1)$-colored graph can be c-triangulated only if it is a partial k-tree. Consider a biconnected 3-colored graph $G = (V, E)$. In order that G can be c-triangulated, G must be a biconnected partial 2-tree.

Theorem 4.2.1 *Let G be a biconnected 3-colored partial 2-tree, and let $\overline{G}$ be the cell completion of G. G can be c-triangulated if and only if the following two conditions hold:*

1. *no two adjacent vertices of $\overline{G}$ have the same color, and*

2. *every cell of $\overline{G}$ has at least three vertices with different colors.*

Proof. From Theorem 4.1.1 it follows that it is sufficient to show that a chordless cycle can be c-triangulated if and only if it contains three vertices of different colors. This fact was proved in [93]. For reasons of completeness we also present a proof here.

If a cycle has only two colors, then it is easy to see that it cannot be c-triangulated, since cycles of length 3 cannot be made. Suppose a cycle S has three colors. Notice that S must have a vertex such that its two neighbors have different colors.

If S is a triangle there is nothing to prove. If S has more than three vertices we can add a chord to S such that the two new cycles made by this chord each have three colors. To find such a chord, we consider two cases. In case that there is a color that appears only once, then take any chord containing this color at one of its end-vertices. If every color appears at least twice in S, then take any vertex v such that its two neighbors have different colors. Then take the chord connecting the two neighbors of v.

This proves that any chordless cycle which is colored by at least three colors can be c-triangulated. □

Remark. Notice that the theorem is not true for t-colored partial 2-trees with $t > 3$. As a counterexample, take $G = K(2, 3)$, which is a partial 2-tree. Let A be the color class with two vertices and B be the color class with three vertices. In the cell-completion of G, A is made an edge. Now color the two vertices of A with the same color a, and the three vertices of B with different colors b_1, b_2 and b_3 with $b_i \neq a$ for $i = 1, 2, 3$. Then the cell-completion is not correctly colored. But G can be c-triangulated by making a clique of B.

A slightly different characterization is obtained in the following theorem.

Theorem 4.2.2 *Let G be a biconnected 3-colored partial 2-tree, and let*
$\overline{G}$ *be the cell completion of G. G can be c-triangulated, if and only if:*

1. *no two adjacent vertices of $\overline{G}$ have the same color, and*

2. *every cycle in $\overline{G}$ contains at least three vertices with different colors.*

Proof. Use Theorem 4.2.1. Clearly, if each cycle in $\overline{G}$ contains at least three different colors, then each cell in the cell completion does so too. This shows the 'if'-part. The 'only if'-part follows from the fact that, if $\overline{G}$ contains a cycle with only two colors, then $\overline{G}$ contains a subgraph that cannot be triangulated, hence G cannot be triangulated.   □

It follows that a 3-colored partial 2-tree can be c-triangulated if and only if $\overline{G}$ does not contain an edge between vertices with the same color, and for every pair of colors, the subgraph of $\overline{G}$ induced by the vertices with these colors is cycle-free, i.e., is a partial 1-tree. This partly generalizes:

Lemma 4.2.2 *Let H be a k-tree colored with $k + 1$ colors such that no two adjacent vertices have the same color. Let S be a subset of the colors and let $W = W(S)$ be the set of vertices with a color in S. Then the induced subgraph $H[W]$ is a $(|S| - 1)$-tree.*

Proof. Let $\sigma = [v_1, \ldots, v_n]$ be a perfect elimination scheme for H. Remove from σ all vertices with a color that is not in S and let σ' be the ordering of the vertices of $H[W]$ obtained in this way. We show that:

1. $|\{v_{n-k}, \ldots, v_n\} \cap W| = |S|$,

2. for all $i \leq n - k$: $|N(v_i) \cap \{v_i, \ldots, v_n\} \cap W| = \begin{cases} |S| - 1 & \text{if } v_i \in W \\ |S| & \text{if } v_i \notin W \end{cases}$

The first item follows from the fact that $\{v_{n-k}, \ldots, v_n\}$ is a $(k + 1)$-clique, and hence each color of S occurs exactly once. The second item follows by a similar argument. It follows that σ' is a perfect elimination scheme for $H[W]$ such that all $v_i \in W$ are adjacent to a clique with $|S| - 1$ vertices in $N(v_i) \cap \{v_i, \ldots, v_n\} \cap W$. This proves the lemma.   □

Corollary 4.2.1 *If a t-colored graph G can be c-triangulated, then for every subset of $s \leq t$ colors, the subgraph of G induced by the vertices with a color in this set is a partial $(s - 1)$-tree.*

Remark. This necessary condition is unfortunately not sufficient, not even for $t = 3$. Construct a graph G as follows. Take $K(2, 3)$. Let A be the color class with two vertices and let B be the color class with three vertices. On every edge put an extra vertex. Let C be the set of these extra vertices. This completes the construction of G. Clearly G is a partial 2-tree since the cell-completion, obtained by creating an edge between the two vertices of A, is a tree of cycles. On the other hand, the induced subgraphs $G[A \cup B]$, $G[A \cup C]$ and $G[B \cup C]$ are all acyclic.

The following result follows directly from Lemma 4.2.1 and Theorem 4.2.2.

Corollary 4.2.2 *Let G be a biconnected 3-colored graph and let $\overline{G}$ be the cell completion of G. G can be c-triangulated if and only if:*

1. *G is a partial 2-tree,*

2. *no two adjacent vertices of $\overline{G}$ have the same color,*

3. *every cycle in $\overline{G}$ contains at least three vertices with different colors.*

4.3   Algorithm for 3-colored graphs

In this section we describe an algorithm to c-triangulate a 3-colored graph whenever such a triangulation is possible. In section 4.4 we give an easier variant that only tests whether it is possible to c-triangulate the graph, without actually yielding a c-triangulation.

Let G be a biconnected 3-colored graph. Recall that a c-triangulation (if it exists) can be obtained by triangulating each chordless cycle of the cell-completion (Theorem 4.1.1). Our algorithm to c-triangulate G has the following structure:

1. Make *any* triangulation of G into a 2-tree H. If this step fails (i.e., G is not a partial 2-tree) then STOP; the graph cannot be c-triangulated.

2. Given H we then can make the cell-completion $\overline{G}$ of G using Lemma 4.1.2. Check if no two adjacent vertices in $\overline{G}$ have the same color. If there are adjacent vertices in $\overline{G}$ which have the same color then STOP; the graph cannot be c-triangulated.

3. Make a list of all cells.

4. Check if all cells have three different colors. If there is a cell with only two different colors then STOP; the graph cannot be c-triangulated. Otherwise, c-Triangulate each cell.

Notice that when step 1, 2 or 4 fails, the graph can not be c-triangulated. Otherwise the algorithm outputs a correct c-triangulation. The correctness of the algorithm follows from Theorem 4.2.1.

Below we describe each step of this algorithm in more detail. As each step has a linear time implementation, we get the following result.

Theorem 4.3.1 *There exists a linear time algorithm that, given a 3-colored graph* G, *tests whether* G *can be c-triangulated and, if so, outputs a c-triangulation of* G.

4.3.1   Triangulating G into a 2-tree H

In this subsection we consider the problem to determine whether there exists a 2-tree H that contains a given graph G as a subgraph and to output such a 2-tree when it exists. It is well-known that this problem can be solved in linear time, see e.g. [124]. For completeness we describe the algorithm also here.

The following lemma is well-known, see for example [171].

Lemma 4.3.1 *If* G *is a biconnected partial 2-tree it can be triangulated into a 2-tree* H *by repeatedly choosing a vertex of degree two, making the neighbors of this vertex adjacent and removing the vertex from the graph. The order in which the vertices are removed form a perfect elimination scheme for* H.

We can implement this as follows. Assume that G is represented by means of adjacency lists. Assume that with each edge (x, y) in the adjacency list of x, there is a pointer to the edge (y, x) in the adjacency list of y. Keep a list of vertices of degree 2.

Step 1 Initialize H to G (i.e., make a copy of the adjacency lists).

Step 2 Take a vertex x of the list of vertices of degree 2. Let y and z be the neighbors of x. Add z to the adjacency list of y and y to the adjacency list of z. It is possible, that a duplicate edge (y, z) s created in this way; in that case y appears twice in the adjacency list of z, and vice versa.

Step 3 Now test, whether y or z has degree 2 (and hence whether they must be put on the list of vertices of degree 2).

Step 4 Scan the adjacency list of y until either we have encountered *three different* neighbors of y, or the list becomes empty. When a duplicate edge is encountered while scanning the list, say (y, x'), remove the second copy of it from the adjacency list of y and remove its counterpart from the list of x'. If y has only two different neighbors put y in the list of vertices of degree two. We do the same for z.

Step 5 Iterate Step 2, 3 and 4 until there are no vertices of degree 2 left. If the graph now has more than two vertices, then STOP; G is not a partial 2-tree. Otherwise, H is a 2-tree, containing G as a subgraph. The order in which the vertices are removed, is a perfect elimination scheme for the 2-tree H.

The correctness of the algorithm follows from [171]. To see that this algorithm runs in linear time notice the following. When scanning the adjacency lists, in every step we either encounter a duplicate edge (which is then removed) or we find a new neighbor. The total number of duplicate edges is at most n (the number of vertices of G), since every time a vertex is removed at most one duplicate edge is created.

We remark here that the description given in [171] is insufficient to show the linear time and linear space bound of the algorithm. In particular, in [171] the test 'given vertices x and y, test whether $(x, y) \in E$' is assumed to take constant time. However, when the edges are given as adjacency lists, this test can take more time (when using a standard model of computation).

4.3.2   Making the cell-completion $\overline{G}$

Suppose that we now have a triangulation into a 2-tree H with a perfect elimination scheme $\sigma = [v_1, \ldots, v_n]$ for H. Recall that $\overline{G}$ is a subgraph of H, so to find $\overline{G}$, we only have to test for each edge in H whether it belongs to $\overline{G}$ or not. We use Lemma 4.1.2 to make the cell-completion. The algorithm is as follows. Make a copy of G in $\overline{G}$. If vertices x and y have at least three common neighbors in H, add the edge (x, y) to $\overline{G}$. This property can be tested as follows.

Step 1 From the adjacency list of a vertex v_i in H, remove the vertices v_j for all $j < i$. Do the same for $\overline{G}$, using the ordering of vertices of σ. Each adjacency list now has at most two elements, since σ is a perfect elimination scheme of a 2-tree.

Step 2 Number the edges of H in any order $1, \ldots, 2n - 3$, and let a pointer point from the edges in the adjacency lists to its number and vice versa. Initialize an array $cn(1 \ldots 2n - 3)$ to zero.

Step 3 Start with the triangle $\{v_{n-2}, v_{n-1}, v_n\}$. For each edge of this triangle, look up its number and increase the value in cn at this position by 1.

Step 4 Consider the other vertices one by one, in the reversed order of the perfect elimination scheme, i.e., in the order $v_{n-3}, v_{n-4}, \ldots, v_1$. For each vertex v_i ($i < n - 2$) do the following. Suppose v_i is adjacent to vertices v_j and v_k (with $j > i$ and $k > i$). For the edges in this triangle, increase the value in cn by one.

Step 5 If a final value in cn is at least three, look up the edge in the adjacency list of H and add this edge to $\overline{G}$ if it was not already present.

Clearly, this procedure uses linear time. It is straightforward to see (by induction) that in Step 4, after vertex v_i has been encountered, for each edge in the induced subgraph $H[\{v_i, v_{i+1}, \ldots, v_n\}]$ the number of common neighbors of its end-vertices is given by the value in cn. We use here that each triangle v_i, v_j, v_k is considered exactly once, namely when considering the lowest numbered vertex in the triangle. Correctness follows now from Lemma 4.1.2.

4.3.3   Making a list of the cells

Notice that the number of cells in $\overline{G}$ is at most $n - 2$ (with equality if and only if $\overline{G}$ is a 2-tree). For each cell initialize an empty list. During the algorithm keep, for each edge in H that has been encountered, pointers to the number of the last cell the edge is contained in and to the positions of its end vertices in this cell.

Again let $\sigma = [v_1, \ldots, v_n]$ be a perfect elimination scheme for H. Recall that for each vertex v_i, all vertices v_j with $j < i$ are removed from its adjacency list.

Step 1 Put the vertices v_n, v_{n-1}, v_{n-2} in the first cell and for each edge in this triangle make pointers to the number of this cell and to the positions of its end vertices in this cell.

Step 2 Consider the other vertices one by one, in the reversed order of the perfect elimination scheme. Suppose v_i has neighbors v_j and v_k in H with $j > i$ and $k > i$. Let $j < k$. Look in the adjacency list of v_j in $\overline{G}$ (which has at most two elements) whether the vertices v_j and v_k are adjacent.

If v_j and v_k are adjacent in $\overline{G}$, make a new cell containing the three vertices v_i, v_j and v_k and for each edge of this triangle update the number of the cell it is contained in.

If v_j and v_k are not adjacent in $\overline{G}$, then they can be contained in at most one cell, since otherwise v_j and v_k would have at least three different neighbors in H (v_i included) and hence would be adjacent in $\overline{G}$. There is a pointer to this cell and to the position of v_j and v_k in this cell. Add the vertex v_i to this cell in the place between vertices v_j and v_k and update the cell number for the edges (v_i, v_j) and (v_i, v_k).

It is straightforward to see by induction that at each step each cell contains the vertices of the cell of $\overline{G}$, restricted to the vertices $v_i, v_{i+1}, \ldots, v_n$. So in this way we obtain a list of all cells in $\overline{G}$ in linear time.

4.3.4 Triangulating each cell

Thus far for each cell a list of vertices is made such that consecutive vertices in this list are adjacent in $\overline{G}$. For each cell do the following.

Step 1 If there are two adjacent vertices in the cell with the same color then STOP; G cannot be c-triangulated.

Step 2 For each of the three colors make a list of vertices in the cell that are of this color, and for every vertex make a pointer to its position in the list containing the vertex.

Step 3 If there is a color with an empty list, the cell can not be triangulated; STOP.

Step 4 Assume every color occurs in the cell. If there is a color such that there is only one vertex of that color, make this vertex adjacent to all other vertices. Otherwise, by going around the cycle clockwise, find a vertex x such that the two neighbors are of a different color. Add the edge between the two neighbors and delete x from the cell and from the color list. Move counterclockwise to the neighbor of x, and start again.

It is easy to see that this algorithm runs in linear time.

4.4 Some variants

A simple variant of the algorithm is the following.

Lemma 4.4.1 *There exists a linear time algorithm that, given a* t*-colored graph* G, *tests whether* G *can be* c*-triangulated into a 2-tree.*

Proof. If $t < 3$, then the graph can not be c-triangulated into a 2-tree. For $t \geq 3$ the same algorithm as used in section 4.3 will do. □

We now describe an algorithm that, given a 3-colored graph G, decides whether G can be c-triangulated. It does not produce a c-triangulation of G. However, it is slightly easier than the algorithm in section 4.3. The algorithm has the following structure:

1. Make *any* triangulation of G into a 2-tree H.

2. Given H make the cell-completion of G, $\overline{G}$. Check that no two adjacent vertices in $\overline{G}$ have the same color.

3. For each pair of colors c_1, c_2, take the subgraph of G induced by vertices of color c_1, or c_2, and check whether this graph does not contain a cycle.

G can be c-triangulated if and only if step 1 succeeds (G is a partial 2-tree), in step 2 no pair of adjacent vertices with the same color is found, and in step 3 all three considered induced subgraphs are cycle-free. The correctness of this procedure follows from Corollary 4.2.2. Steps 1 and 2 can be implemented as in section 4.3.1 and section 4.3.2. It is easy to see, that step 3 has a linear time implementation.

Chapter 5

Only few graphs have bounded treewidth

In this chapter we consider the treewidth of random graphs. To be more precise, for random graphs with n vertices and $m \geq \delta n$ edges, we obtain asymptotic lower bounds for the treewidth. If $\delta \geq 1.18$, then almost every (a.e.) graph with $m \geq \delta n$ edges has treewidth $\Theta(n)$. Our results also show the following. Let $\mathcal{G}$ be a class of graphs which is closed under taking minors. Let $\delta \geq 1.18$. Then almost every graph with n vertices and m edges, where $m \geq \delta n$, is *not* an element of $\mathcal{G}$. Part of our results, concerning clique minors of graphs, extend earlier results of Bollobás, Catlin and Erdös [34] and is related to results of Kostochka [112] and Thomason [164].

When looking for bounds for the treewidth of random graphs it is a natural approach to look for subgraphs or minors with large treewidth. The following two results show that this approach is only of limited use (at least for the grid subgraph and the clique minor).

In a recent paper [168], de la Vega and Manoussakis obtain the following related result. They show that for $\epsilon < 1$, the random graph $G_{n,p}$ with edge probability $p = ((1 + \epsilon) \log n / n)^{1/2}$ contains a square $\sqrt{\frac{\epsilon n}{2}} \times \sqrt{\frac{\epsilon n}{2}}$-grid, implying that such a random graph has treewidth $\Omega(\sqrt{n})$.

Bollobás et al. [34] obtain the following result. Let $0 < p < 1$ be fixed. For a.e. graph $G_{n,p}$ the maximum value s such that $G_{n,p}$ has a minor K_s is $(1 + o(1)) \frac{n}{\sqrt{\log_d n}}$, where $d = \frac{1}{q}$.

A related result in extremal graph theory is the following. For a graph G, let $e(G)$ be the number of edges and let $|G|$ the number of vertices. For graphs G and H let $G > H$ denote the 'G has H as a minor' relation. Let $c(s) = \inf\{c \mid e(G) \geq c|G| \Rightarrow G > K_s\}$. The result of [34] shows that $c(s) \geq 0.265 s \sqrt{\log_2 s}$ for large values of s (see [164]). Subsequently, Kostochka [112] showed that $s\sqrt{\log s}$ is the correct order for $c(s)$. Thomason [164] shows the best upper bound as far as we know: $c(s) \leq 2.68 s \sqrt{\log_2 s}(1 + o(1))$, for large

s. For the treewidth problem this bound is of no interest; although a graph with K_s as a minor has treewidth at least $s - 1$, a graph with treewidth at most s, can have at most $ns - \frac{1}{2}s(s + 1)$ edges.

We extend the result of [34] as follows. Let $\delta > 1.18$. Then there exists a positive constant c_δ such that a.e. graph $G_{n,m}$ with $m \geq \delta n$ has a minor K_s with $s \geq \lfloor c_\delta^{2/3} n^{1/3} \rfloor$.

The most important method we use in this chapter is similar to that used in [167] in relation to the *bandwidth* of random graphs. In [167] de la Véga proves that almost all graphs on n vertices with cn edges have bandwidth $\geq b_c n$ where b_c is strictly positive for $c > 1$.

We show that if $\delta \geq 1.18$, almost all graphs with n vertices and δn edges have treewidth $\geq b_\delta n$, where b_δ is strictly positive.

Finally we like to mention two more related results. Cohen et al. [46] give the exact asymptotic probability that a graph is an interval graph and that a graph is a circular arc graph (this paper also contains numerous applications of interval graphs). In the common random graph model, interval graphs play only a minor role. (If $m/n^{5/6} \to \infty$, where m is the number of edges and n is the number of vertices of a random graph, then the probability that the random graph is an interval graph goes to zero, if n tends to infinity.) For this reason, some work has been done to find separate models for random interval graphs, see e.g. [151]. This paper of Scheinerman also contains results on maximum degree, Hamiltonicity, chromatic number etc., for random interval graphs. In this chapter we only look at the common random graph model, i.e., all graphs with a certain number of vertices and a certain number of edges are equi-probable.

We end this overview with the following, hardly related, result which is nevertheless of interest. In [58] it is shown among other things that every graph with n vertices and $m = \frac{n^2}{4} + 1$ edges has a triangulated *subgraph* with $\frac{3n}{2} - 1$ edges.

We summarize the results of this chapter as follows:

1. In section 5.1 we start with restating a result of Bollobás: almost all graphs with $< \frac{1}{2}n$ edges have treewidth ≤ 2.

2. In section 5.2 we show that for all $\delta > 1$ and for all $0 < \epsilon < (\delta - 1)/(\delta + 1)$, a.e. graph $G_{n,m}$ with $m \geq \delta n$ has treewidth $\geq n^\epsilon$.

3. In section 5.3, we show that:

   (a) For all $0 < b < 1$ there exists a constant δ such that if $m \geq \delta n$, then a.e. graph $G_{n,m}$ has treewidth $\geq bn$.

   (b) For all $\delta \geq 1.18$, if $m \geq \delta n$, then a.e. graph $G_{n,m}$ has treewidth $\Theta(n)$.

4. In section 5.4 we prove the following related results:

(a) Let $\delta \geq 1.18$. Then a.e. graph $G_{n,m}$ with $m \geq \delta n$ has a clique minor K_s with $s = \Omega(n^{1/3})$.

(b) Let $\mathcal{G}$ be a minor-closed class of graphs. For all $\delta \geq 1.18$, a.e. graph $G_{n,m}$ with $m \geq \delta n$ is *not* in $\mathcal{G}$.

We do not know whether these results are optimal. Indeed, the smallest constant c such that almost every graph with cn edges has treewidth bn, for some $b > 0$, remains an open problem. Also, we do not have any results on random graphs of which the number of edges is in the range $(\frac{1}{2}n, n)$.

5.1 Preliminaries

For random graphs in general, the reader is referred to [33]. Let $P(Q)$ be the probability that a random graph with n vertices and m edges has a certain property Q. In most cases m is a function of n. We say that almost every (a.e.) graph has property Q if $P(Q) \to 1$ as $n \to \infty$. Throughout this chapter we use $N = \binom{n}{2}$, where n is the number of vertices of a graph. (N is the number of edges in the complete graph K_n with n vertices.) We use $G_{n,p} \in \mathcal{G}(n, p)$ to denote a random graph with edge probability p, and $G_{n,m} \in \mathcal{G}(n, m)$ for a random graph with n vertices and m edges. In [33] it is shown that if m is close to $pN = p\binom{n}{2}$, the two models $\mathcal{G}(n, m)$ and $\mathcal{G}(n, p)$ are practically interchangeable. In fact, since treewidth $\geq k$ is a *monotone increasing* property, it follows that, if $pqN \to \infty$ and x is some fixed constant, almost every graph in $\mathcal{G}(n, p)$ has treewidth $\geq k$ if and only if a.e. graph in $\mathcal{G}(n, m)$ has treewidth $\geq k$, where $m = \lfloor pN + x(pqN)^{\frac{1}{2}} \rfloor$, ([33], page 35).

As a first result on treewidth we can restate a result of Bollobás, ([33], page 99). Define a connected unicyclic graph with t vertices to be any connected graph with t vertices and t edges. Notice that a unicyclic graph has treewidth at most 2.

Lemma 5.1.1 *Suppose* $p = \frac{c}{n}$, $0 < c < 1$. *Then a.e.* $G_{n,p}$ *is such that every connected component is a tree or a unicyclic graph.*

Corollary 5.1.1 *If* $m < \frac{1}{2}n$, *then a.e. graph* $G_{n,m}$ *has treewidth at most two.*

5.2 A bound on the number of subgraphs of k-trees

In this section we show that almost all graphs with δn edges, have treewidth $\geq n^\epsilon$, for all fixed $\epsilon < \frac{\delta - 1}{\delta + 1}$. We do this by proving an upper bound for the number of subgraphs of k-trees.

Recall that the number of k-trees is given by the following formula, which was shown in different manners in a number of papers [10, 63, 99, 128, 140].

$$T_k(n) = \binom{n}{k}(1 + k(n - k))^{n-k-2}$$

Lemma 5.2.1 *Let $0 < \epsilon < 1$, and let $k \le n^\epsilon$. Then for $n \to \infty$:*

$$T_k(n) = o\left(n^{(1+\epsilon)(n-2)}\right)$$

Proof.

$$
\begin{aligned}
T_k(n) &= \binom{n}{k}(1 + k(n - k))^{n-k-2} \le n^k(nk)^{n-k-2} \\
&\le n^{n-2}n^{\epsilon(n-k-2)} = n^{(1+\epsilon)(n-2)}n^{-\epsilon k} = o\left(n^{(1+\epsilon)(n-2)}\right)
\end{aligned}
$$

This proves the lemma.   □

Lemma 5.2.2 *For any integer k, the number of partial k-trees with m edges is at most $T_k(n)\binom{nk}{m}$.*

Proof. The number of edges in a k-tree is $nk - \frac{1}{2}k(k+1)$ (see Lemma 2.1.10 on page 13 or e.g. [10]). It follows that the number of partial k-trees with m edges is at most

$$T_k(n)\binom{nk - \frac{1}{2}k(k+1)}{m} \le T_k(n)\binom{nk}{m}$$

□

Theorem 5.2.1 *Let $\delta > 1$ and $0 < \epsilon < \frac{\delta-1}{\delta+1}$. Then for all $m \ge \delta n$, a.e. $G_{n,m}$ has treewidth at least n^ϵ.*

Proof. Let k be any integer with $k \le n^\epsilon$. The total number of graphs with m edges is $\binom{N}{m}$, where $N = \binom{n}{2}$. Let F be the fraction of all graphs with m edges that have treewidth $\le k$. We show that $F \to 0$. If $m \ge nk$, then clearly $F = 0$. Assume henceforth that $m < nk$. We have, if n is sufficiently large:

$$F \le T_k(n)\frac{\binom{nk}{m}}{\binom{N}{m}} \le T_k(n)\left(\frac{nk}{N-m}\right)^m \le T_k(n)\left(3\frac{n^{1+\epsilon}}{n^2}\right)^{\delta n}$$

since $m < nk \le n^{1+\epsilon}$ and $N - m \ge \frac{1}{3}n^2$ if n is large enough. We also used $m \ge \delta n$ and $k \le n^\epsilon$. It follows that for large enough n (using Lemma 5.2.1):

$$F \le T_k(n)\left(3n^{\epsilon-1}\right)^{\delta n} \le 3^{\delta n}n^{n(\epsilon(\delta+1)-(\delta-1))} \to 0$$

since $\epsilon(\delta + 1) - (\delta - 1) < 0$.   □

In the next section we show that, if δ is somewhat larger, almost every graph with n vertices and at least δn edges has a treewidth which is linear in n.

5.3 The separator method

In this section we show the following. Let $\delta \geq 1.18$. There exists a *positive* number b_δ, such that for $m \geq \delta n$, a.e. graph $G_{n,m}$ does not have a balanced separator of size $\leq b_\delta n$. It is well-known that a partial k-tree has a balanced separator with at most $k+1$ vertices. We show that a random graph with at least δn edges does not have such a separator for small k.

In this section a balanced separator is a set C with $k+1$ vertices such that every connected component has at most $\frac{1}{2}(n-k)$ vertices. Recall Lemma 2.2.9 on page 23 which shows there exist such separators in partial k-trees:

> If G is a k-tree then there is a clique C in G with $k+1$ vertices such that every connected component of $G[V \setminus C]$ has at most $\frac{1}{2}(n-k)$ vertices.

Corollary 5.3.1 *Let $G = (V,E)$ be a graph with at least $k+1$ vertices and with treewidth $\leq k$. Then there exists a set S of $k+1$ vertices such that every component of $G[V-S]$ has at most $\frac{1}{2}(n-k)$ vertices.*

Let $G = (V,E)$ be a graph. To ease the later computations somewhat, we partition the vertices of G in three sets.

Definition 5.3.1 *Let $G = (V,E)$ be a graph with n vertices. A partition (S,A,B) of the vertices is a* balanced k-partition *if the following three conditions are satisfied:*

1. $|S| = k+1$

2. $\frac{1}{3}(n-k-1) \leq |A|, |B| \leq \frac{2}{3}(n-k-1)$

3. *S separates A and B, i.e., there are no edges between vertices of A and vertices of B.*

Lemma 5.3.1 *Let $G = (V,E)$ be a partial k-tree with n vertices such that $n \geq k+4$. Then G has a balanced k-partition.*

Proof. Let S be a balanced separator in G, which exists by Corollary 5.3.1. Let $C_1, \ldots, C_t$ be the connected components of $G[V-S]$. Hence, $|C_i| \leq \frac{1}{2}(n-k)$. We consider two cases:

Case 1: There exists a component C_i such that $|C_i| \geq \frac{1}{3}(n-k-1)$.
In this case let $A = C_i$ and B the union of the other components. Then clearly, since $n-k \geq 4$, and $|C_i| \leq \frac{1}{2}(n-k)$ it follows that $|A| \leq \frac{2}{3}(n-k-1)$.

Case 2: All components have less than $\frac{1}{3}(n-k-1)$ vertices.
In this case, choose s such that:

$$|C_1| + \ldots + |C_s| \leq \frac{2}{3}(n-k-1) < |C_1| + \ldots + |C_{s+1}|$$

It follows that:

$$|C_{s+2}| + \ldots + |C_t| < \frac{1}{3}(n - k - 1)$$

Let $A = C_{s+1} \cup \ldots \cup C_t$ and $B = C_1 \cup \ldots \cup C_s$. Then:

$$|A| < |C_{s+1}| + \frac{1}{3}(n - k - 1) \leq \frac{2}{3}(n - k - 1)$$

Clearly, also:

$$|A| = n - k - 1 - |B| \geq \frac{1}{3}(n - k - 1)$$

$\square$

We show that the fraction of all graphs that have a balanced k-partition is negligible if the number of edges is not too small.

Lemma 5.3.2 *Let $L_k(n, m)$ be the number of graphs with m edges that have a balanced k-partition. Then an upper bound for $L_k(n, m)$ is:*

$$\frac{1}{2} \sum_{\frac{1}{3}(n-k-1) \leq a \leq \frac{2}{3}(n-k-1)} \binom{n}{k+1} \binom{n-k-1}{a} \binom{N - a(n-k-1-a)}{m}$$

where $N = \binom{n}{2}$.

Proof. First choose a separator S with $k + 1$ vertices, and a set A with a vertices, where $\frac{1}{3}(n - k - 1) \leq a \leq \frac{2}{3}(n - k - 1)$. Finally, choose m edges. Since no edges between A and B are allowed, these m edges must be chosen from a set of $N - a(n - k - 1 - a)$ available edges. Since A and B are interchangeable, we can divide by 2 to find an upper bound for the number of graphs which have a balanced k-partition. $\square$

Lemma 5.3.3

$$L_k(n, m) \leq \binom{n}{k+1} \cdot 2^{n-k-2} \cdot \binom{N - \frac{2}{9}(n-k-1)^2}{m}$$

Proof. Consider the bound on $L_k(n, m)$ in Lemma 5.3.2. Notice that:

$$\frac{1}{3}(n-k-1) \leq a \leq \frac{2}{3}(n-k-1) \Rightarrow N - a(n-k-1-a) \leq N - \frac{2}{9}(n-k-1)^2$$

and hence

$$\binom{N - a(n-k-1-a)}{m} \leq \binom{N - \frac{2}{9}(n-k-1)^2}{m}$$

Also notice that

$$\sum_{\frac{1}{3}(n-k-1) \leq a \leq \frac{2}{3}(n-k-1)} \binom{n-k-1}{a} \leq 2^{n-k-1}$$

Using these upper bounds, the lemma follows. $\square$

Definition 5.3.2 *Let $F_k(n, m)$ be the fraction of all graphs with n vertices and m edges that have a balanced k-partition, i.e.,*

$$F_k(n, m) = \frac{L_k(n, m)}{\binom{N}{m}}$$

where $N = \binom{n}{2}$.

Lemma 5.3.4

$$F_k(n, m) \le 2^{n-k-2} \cdot \binom{n}{k+1} \cdot \left(1 - \frac{\frac{4}{9}(n-k-1)^2}{n^2}\right)^m$$

Proof. Let $t = \frac{2}{9}(n-k-1)^2$. Then

$$\frac{\binom{N-t}{m}}{\binom{N}{m}} = \frac{N-t}{N} \cdot \frac{N-t-1}{N-1} \cdot \dots \cdot \frac{N-t-m+1}{N-m+1}$$

$$= \left(1 - \frac{t}{N}\right)\left(1 - \frac{t}{N-1}\right)\dots\left(1 - \frac{t}{N-m+1}\right)$$

$$\le \left(1 - \frac{t}{N}\right)^m \le \left(1 - \frac{2t}{n^2}\right)^m$$

Using this the lemma follows from Lemma 5.3.3. $\qquad\square$

Definition 5.3.3 *For $0 < b < 1$ and $\delta > 0$, let:*

$$\varphi(b, \delta) = 2^{1-b} \cdot \frac{\left(1 - \frac{4}{9}(1-b)^2\right)^\delta}{b^b(1-b)^{1-b}}$$

Theorem 5.3.1 *Let $0 < b < 1$ and δ be fixed. Let $m \ge \delta n$ and let $k + 1 = \lceil bn \rceil$. Then*

$$F_k(n, m) = o\left(\varphi(b, \delta)^n\right)$$

Proof. Since $k + 1 \to \infty$ and $n - k - 1 \to \infty$, we find with the aid of Stirling's formula:

$$\binom{n}{k+1} \sim \left(\frac{n}{k+1}\right)^{k+1} \cdot \left(\frac{n}{n-k-1}\right)^{n-k-1} \cdot \frac{1}{\sqrt{2\pi(k+1)(1 - \frac{k+1}{n})}}$$

Since $bn \le k + 1 \le bn + 1$, it follows that $\left(\frac{n}{k+1}\right)^{k+1} \le \left(\frac{1}{b}\right)^{bn+1}$. And also

$$\left(\frac{n}{n-k-1}\right)^{n-k-1} \le \left(\frac{n}{n-bn-1}\right)^{n-bn}$$

$$= \left(\frac{1}{1-b}\right)^{(1-b)n} \cdot \frac{1}{\left(1 - \frac{1}{n(1-b)}\right)^{n(1-b)}}$$

$$\sim e \cdot \left(\frac{1}{1-b}\right)^{(1-b)n}$$

Since $m \geq \delta n$ we have

$$\left(1 - \frac{\frac{4}{9}(n - k - 1)^2}{n^2}\right)^m \leq \left(1 - \frac{\frac{4}{9}(n - bn - 1)^2}{n^2}\right)^{\delta n}$$

$$\leq \left(1 - \frac{4}{9}(1 - b)^2 + \frac{8}{9n}(1 - b)\right)^{\delta n}$$

$$\leq \left(1 - \frac{4}{9}(1 - b)^2\right)^{\delta n} \cdot \left(1 + \frac{8(1 - b)}{9n(1 - \frac{4}{9}(1 - b)^2)}\right)^{\delta n}$$

$$\sim \left(1 - \frac{4}{9}(1 - b)^2\right)^{\delta n} \cdot \exp\left(\frac{8\delta(1 - b)}{9 - 4(1 - b)^2}\right)$$

The result now follows from the fact that $\sqrt{2\pi(k + 1)}(1 - \frac{k+1}{n}) \to \infty$ and
Lemma 5.3.4.   □

Theorem 5.3.2

1. For all $0 < b < 1$ there exists a δ_b such that if $m \geq \delta_b n$ a.e. graph $G_{n,m}$ has treewidth $\geq bn$.

2. Let $\delta \geq 1.18$. There exists a positive constant b_δ such that if $m \geq \delta n$ and $k + 1 = \lceil b_\delta n \rceil$, then $F_k(n, m) \to 0$.

3. Let $\delta \geq 1.18$. Then a.e. graph $G_{n,m}$ with $m \geq \delta n$ has treewidth $\Theta(n)$.

Proof. Notice that $\lim_{\delta \to \infty} \varphi(b, \delta) = 0$. The first statement now follows from Theorem 5.3.1.

To prove the second statement, notice that

$$\lim_{b \downarrow 0} \left(\frac{1}{b}\right)^b \cdot \left(\frac{2}{1 - b}\right)^{1-b} = 2$$

Hence $\varphi(0, \delta) = 2(\frac{5}{9})^\delta < 1$ if $\delta \geq 1.18$. Hence, from the fact that φ is a continuous function for b in $(0, 1)$ it follows that for $\delta \geq 1.18$, there exists a *positive number* b_δ such that $\varphi(b_\delta, \delta) < 1$. The second statement now follows immediately Theorem 5.3.1.

The third statement is an immediate consequence of the second one.   □

5.4 Further results

In this section, we show some results in the theory of graph minors. To be more specific, we show that, for any minor-closed class of graphs, a.e. graph with at least a certain amount of edges is *not* in the class. Apart from Theorem 5.3.2, the following theorem is the main ingredient. In [1] Alon et al. proved the following theorem.

Theorem 5.4.1 *Let $h \geq 1$ be an integer and let $G = (V, E)$ be a graph with n vertices and no K_h minor. There exists a subset $X \subseteq V$, with $|X| \leq h^{3/2} n^{1/2}$ such that every connected component of $G[V - X]$ has at most $\frac{1}{2} n$ vertices.*

Together with our results this proves the following theorem.

Theorem 5.4.2 *Let $\delta \geq 1.18$. There exists a positive constant c_δ, such that a.e. graph $G_{n,m}$ with $m \geq \delta n$ edges has a clique K_h with $h \geq \lfloor c_\delta n^{1/3} \rfloor$ as a minor.*

Proof. By Theorem 5.3.2, there is a positive number b_δ such that a.e. graph $G_{n,m}$ with $m \geq \delta n$, does not have a balanced k-partition for $k \leq b_\delta n$. Let $c_\delta = (\frac{1}{3} b_\delta)^{2/3}$ and let $h = \lfloor c_\delta n^{1/3} \rfloor$. By Theorem 5.4.1, if a graph does not have K_h as a minor, then it has a separator X with $|X| \leq c_\delta^{3/2} n \leq \frac{1}{3} b_\delta n$, such that every component has at most $\frac{1}{2} n$ elements. By a similar argument as in Lemma 5.3.1 there is a partition of the vertices (X, A, B) such that $|A|, |B| \leq \frac{2}{3} n$. Assume this is not a balanced k-partition. Then without loss of generality we may assume that $|A| > \frac{2}{3}(n - |X|)$. Let $t = 3|A| - 2(n - |X|)$. Take t vertices out of A and move them into X. Call the new sets X' and A'. Clearly, X' separates A' and B, and $|A'| = \frac{2}{3}(n - |X'|)$. Since $|A| \leq \frac{2}{3} n$:

$$|X'| = |X| + t = |X| + 3|A| - 2(n - |X|) \leq 3|X|$$

It follows that (X', A', B) is a balanced k-partition with $k + 1 = |X'| \leq b_\delta n$. □

Let $\mathcal{G}$ be a minor-closed class of graphs (e.g. the class of planar graphs). Robertson and Seymour [142] proved Wagner's conjecture; there is a *finite* set of forbidden minors. From this result the following lemma easily follows.

Lemma 5.4.1 *Let $\mathcal{G}$ be a minor-closed class of graphs. Then there exists an integer h such that K_h is not a minor of any graph $G \in \mathcal{G}$.*

Proof. Take a finite set S of forbidden minors. Let h be the minimum number of vertices of any element of S. □

Theorem 5.4.3 *Let $\mathcal{G}$ be a minor-closed class of graphs. For all $\delta \geq 1.18$, a.e. graph $G_{n,m}$ with $m \geq \delta n$ is not in $\mathcal{G}$.*

Proof. Take h such that no graph in $\mathcal{G}$ has K_h as a minor. From Theorem 5.4.2, there exists a positive number c_δ, such that a.e. graph $G_{n,m}$ with $m \geq \delta n$ has a K_t minor with $t \geq \lfloor c_\delta n^{1/3} \rfloor$. It immediately follows that a.e. graph $G_{n,m}$ is not in $\mathcal{G}$. $\square$

Chapter 6

Approximating treewidth and pathwidth of a graph

In this chapter we look at the problem of efficiently approximating treewidth and pathwidth for graphs in general. Since computing the treewidth or pathwidth of a graph is NP-hard, it is of interest to have a good approximation algorithm. In this chapter, we show first that there exists a fast approximation algorithm for the treewidth of a graph with performance ratio $O(\sqrt{kn})$, where n is the number of vertices of the graph and k is the treewidth. We also give a polynomial time approximation algorithm with performance ratio $O(\log n)$.

In fact, the first algorithm we give computes a *path-decomposition* with width at most $O(k^{\frac{3}{2}}\sqrt{n})$ where k is the *treewidth* and n the number of vertices of the graph. We give an upper bound for the *exact* performance ratio, which shows that the constant is quite reasonable. Our main tool is the algorithm of Alon et al. [1], for finding good vertex separators. We use this algorithm to find vertex separators, because of its simplicity, and nice behavior (i.e. there are no large hidden constants involved). It is interesting to notice that if G is a *planar* graph then our algorithm computes a path-decomposition of G of width $O(\sqrt{n})$.

Recently, Seymour and Thomas [153] have obtained a polynomial algorithm for the related notion of *branchwidth*, restricted to planar graphs. As the branchwidth and treewidth of a graph differ by at most a factor of $\frac{3}{2}$ [145], this gives a polynomial time approximation algorithm for treewidth of planar graphs with performance ratio $\frac{3}{2}$, and polynomial time approximation algorithms for pathwidth of planar graphs with performance ratio $O(\log n)$. Interesting questions are whether a polynomial time algorithm exists that determines treewidth exactly for planar graphs, and whether the pathwidth of a planar graph can be approximated with a constant performance ratio.

Our methods show that basically *any* approximation algorithm for balanced vertex separators can be used to find approximations for the treewidth and pathwidth.

Recently, it was reported in [95] that Leighton and Rao obtain balanced vertex separators with performance ratio $O(\log n)$, using an approximate max-flow min-cut theorem [119]. However, we do not know how fast the algorithm is and neither do we know much about the constant in this approximation. However, we do show that such an algorithm can be used to give an approximation algorithm for the treewidth of a graph with performance ratio $O(\log n)$. With a simple trick we can then find a path-decomposition of which the width is at most factor $O(\log^2 n)$ off the optimal width.

To our knowledge it is open whether the problem of approximating the treewidth within a *constant* factor is NP-hard or not. In fact, we feel this is one of the biggest open problems in the research dealing with treewidth and pathwidth at the moment. If the fast algorithms for solving NP-hard problems for graphs with bounded treewidth are ever to become of practical importance, it is undoubtedly of importance to find good tree-decompositions for these graphs (having small width). Since the $O(\log n)$ approximations do not seem to make many of these algorithms practical, it is of great interest to know whether approximations within a small constant exist. We show in this chapter that finding approximations within an *additive* constant (absolute approximation) is NP-hard.

6.1 Approximating pathwidth

In this section we give a polynomial algorithm to approximate the pathwidth of a graph. We use the algorithm of Alon et al. (Theorem 5.4.1, page 59) to find balanced vertex separators. For convenience we recall their result (using our own terminology). As usual, if $G = (V, E)$ is a graph and $w : V \rightarrow \mathbf{R}^+$ is a weight function which maps every vertex onto a non-negative real number, then for a subset Q of the vertices, we denote by $w(Q)$ the sum of the weights of the vertices in Q.

> There is an algorithm with running time $O(m\sqrt{hn})$, which takes as input an integer $h \geq 1$, a graph $G = (V, E)$ (with n vertices and $m = |V| + |E|$), and a weight function $w : V \rightarrow \mathbf{R}^+$ (the set of non-negative real numbers), and that outputs either:
>
> - a K_h-minor of G, or
> - a subset X of V, with $|X| \leq h\sqrt{hn}$ such that $w(F) \leq \frac{1}{2}w(V)$ for every connected component F of $G[V \setminus X]$.

We need only a restricted version of the theorem of Alon et al. where the weights of the vertices are restricted to values of zero and one.

Definition 6.1.1 *Let* $G = (V, E)$ *be a graph,* $W \subseteq V$ *a subset of the vertices and* α *any real number between 0 and 1. An* α-*vertex-separator of* W *in* G *is a set of vertices* $S \subseteq V$ *such that every connected component of* $G[V \setminus S]$ *has at most* $\alpha |W|$ *vertices of* W.

The following lemma is an immediate consequence of the theorem of Alon et al.

Lemma 6.1.1 *There is an algorithm with running time* $O((kn)^{\frac{3}{2}})$, *which takes as input an integer* k, *a graph* $G = (V, E)$ *(with* $|V| = n$*), and a subset of the vertices* Q *and that outputs either:*

- *a valid statement that the treewidth of* G *is larger than* k, *or*

- *a* $\frac{1}{2}$-*vertex-separator* $X \subseteq V$ *of* Q *in* G *with* $|X| \leq (k+2)^{\frac{3}{2}} \sqrt{n}$.

Proof. First check if the number of edges in G does not exceed $\frac{1}{2}k(2n - k - 1)$. If it does, the graph cannot have treewidth $\leq k$. Otherwise use the theorem of Alon et al., with $h = k + 2$, and a 0/1 valued weight function which has value 1 precisely for the vertices of Q. Notice that if K_h is a minor of G, then the treewidth is at least $h - 1$. □

Definition 6.1.2 *Let* $G = (V, E)$ *be a graph with* n *vertices. A* partition *of* G *is a partition* (X, A, B) *of the set of vertices into three subsets such that:*

- X *separates* A *and* B, *and*

- $|A| \leq \frac{2}{3}n$, $|B| \leq \frac{2}{3}n$.

Lemma 6.1.2 *There is an algorithm with running time* $O((nk)^{\frac{3}{2}})$ *which takes as input an integer* k *and a graph* $G = (V, E)$. *It outputs either:*

- *a correct statement that the treewidth of* G *is larger than* k, *or*

- *a partition* (X, A, B) *with* $|X| \leq (k+2)^{\frac{3}{2}} \sqrt{n}$.

Proof. Use Lemma 6.1.1 to either determine that the treewidth of G exceeds k or find an $\frac{1}{2}$-vertex-separator X of V, such that $|X| \leq (k+2)^{\frac{3}{2}} \sqrt{n}$. Let $\frac{1}{2} \leq \alpha \leq \frac{2}{3}$. Notice that X is an α-vertex-separator with $|X| \leq (k+2)^{\frac{3}{2}} \sqrt{n}$. Let $C_1, C_2, \ldots, C_t$ be the vertex sets of the connected components of $G[V \setminus X]$. We group these sets together into two sets A and B as follows.

- If there is a set C_i with $|C_i| \geq \frac{1}{3}n$ then take $A = C_i$ and $B = V \setminus (X \cup A)$.

- If $|V \setminus X| \leq \frac{2}{3}n$, then take $A = V \setminus X$ and $B = \emptyset$.

- In all other cases take ℓ such that $|C_1| + \ldots + |C_{\ell-1}| < \frac{1}{3}n$ and $|C_1| + \ldots + |C_\ell| \geq \frac{1}{3}n$. Take $A = C_1 \cup \ldots \cup C_\ell$ and $B = V \setminus (X \cup A)$.

This partitions the vertices of $V \setminus X$ into two sets A and B with $|A|, |B| \leq \frac{2}{3}n$, and such that X separates A and B. $\square$

In figure 6.1 we present the algorithm to obtain a path-decomposition.

procedure Pathdec(G, k, P);
comment {*Input: Graph* $G = (V, E)$ *with* n *vertices and integer* k.
        Output: A path-decomposition P *of* G *or*
        a correct statement that the treewidth of G *is larger than* k}.
begin
  $\beta \leftarrow (1/(1 - \sqrt{\frac{2}{3}}))^2$;
  if $n \leq \beta(k + 2)^3$ **then**
    let P consist of one subset containing all vertices of G
  else
    begin
      determine whether there exists a partition (X, A, B) in G with
      $|X| \leq (k + 2)^{\frac{1}{2}}\sqrt{n}$ (Lemma 6.1.2);
      if such a partition is not found **then** output("the treewidth exceeds k")
      else
        begin
          call pathdec$(G[A], k, P_A)$;
          call pathdec$(G[B], k, P_B)$;
          add X to all subsets of P_A and to all subsets of P_B;
          let P'_A and P'_B be the new path-decompositions;
          $P \leftarrow P'_A \mathbin{+\!\!+} P'_B$
        end
    end
end

Figure 6.1: Procedure to find a path-decomposition.

First we show that the algorithm is correct.

Lemma 6.1.3 *Let* $G = (V, E)$ *be a graph with* n *vertices, and let* k *be an integer.* Pathdec(G, k, P) *either outputs a correct statement saying that the treewidth of* G *exceeds* k *or returns a path-decomposition* P *of* G.

Proof. Let $\beta = (1/(1 - \sqrt{\frac{2}{3}}))^2$. Clearly, if the algorithm outputs that the treewidth exceeds k, then this is indeed the case by Lemma 6.1.2. Assume this does not occur. We show that the algorithm outputs a path-decomposition P. Clearly this is the case if $n \leq \beta(k + 2)^3$. Let

$n > \beta(k+2)^3$. Assume that the recursive calls $\text{pathdec}(G[A], k, P_A)$ and $\text{pathdec}(G[B], k, P_B)$ return path-decompositions P_A of $G[A]$ and P_B of $G[B]$ (if not, the treewidth of G exceeds k). In the next step all vertices of X are added to all subsets of the path-decompositions P_A and P_B, thus creating P'_A and P'_B. Clearly P'_A and P'_B are correct path-decompositions for $G[A \cup X]$ and $G[B \cup X]$. Finally, it easily follows that $P = P'_A ++ P'_B$ is a path-decomposition of G. □

Lemma 6.1.4 *If G is a graph with n vertices and with treewidth k, then $\text{pathdec}(G, k, P)$ returns a path-decomposition P of G of width at most*

$$\frac{1}{1 - \sqrt{\frac{2}{3}}}(k+2)^{\frac{3}{2}}\sqrt{n} - 1.$$

Proof. Let $\beta = (1/(1 - \sqrt{\frac{2}{3}}))^2$. If $n \leq \beta(k+2)^3$, the result follows directly. Henceforth assume without loss of generality that $n > \beta(k+2)^3$.

Let $\alpha = \frac{1}{1 - \sqrt{\frac{2}{3}}}(k+2)^{\frac{3}{2}}$.

Now, by induction, the recursive calls of pathdec for $G[A]$ and $G[B]$ return path-decompositions P_A of $G[A]$ and P_B of $G[B]$ of width at most $\alpha\sqrt{|A|} - 1$ and $\alpha\sqrt{|B|} - 1$, respectively. Hence we find for the width of P, since $|A| \leq \frac{2}{3}n$ and $|B| \leq \frac{2}{3}n$:

$$\begin{aligned}
\text{width}(P) &\leq (k+2)^{\frac{3}{2}}\sqrt{n} + \alpha \max(\sqrt{|A|}, \sqrt{|B|}) - 1 \\
&\leq (k+2)^{\frac{3}{2}}\sqrt{n} + \alpha\sqrt{\frac{2}{3}n} - 1 = \alpha\sqrt{n} - 1.
\end{aligned}$$

□

We can now state the main result.

Theorem 6.1.1 *There is an algorithm with running time $O((nk)^{\frac{3}{2}})$ which takes as input an integer k and a graph $G = (V, E)$ with n vertices and that outputs either:*

- *a correct statement that the treewidth of G exceeds k, or*

- *a path-decomposition of G of width at most*

$$\frac{1}{1 - \sqrt{\frac{2}{3}}}(k+2)^{\frac{3}{2}}\sqrt{n} - 1.$$

Proof. By Lemma 6.1.3 and Lemma 6.1.4 it remains to verify the running time of the algorithm. First we show that the number of subsets returned in the path-decomposition for a graph with n vertices is at most n. This is clearly the case when $n \leq \beta(k+2)^3$. Otherwise the number of subsets is the number of subsets in P_A plus the number of subsets in P_B, which is at most $|A| + |B| \leq n$ (by induction).

Let $T_k(n)$ be the running time of $pathdec(G, k, P)$. We show that there is a constant C such that $T_k(n) \leq C(k+2)^{\frac{3}{2}} n^{\frac{3}{2}}$. This is clearly so when $n \leq \beta(k+2)^3$. Otherwise, finding the partition takes time $O((nk)^{\frac{3}{2}})$, by Lemma 6.1.2. By induction, the recursive calls can be done in $C(k+2)^{\frac{3}{2}}(|A|^{\frac{3}{2}} + |B|^{\frac{3}{2}})$ time. Finally adding all vertices of X to the subsets of P_A and P_B also takes time $O((nk)^{\frac{3}{2}})$, since $|X| \leq (k+2)^{\frac{3}{2}} \sqrt{n}$ and the total number of subsets is at most n. Thus we find for some constant γ:

$$T_k(n) \leq C(k+2)^{\frac{3}{2}} \left(|A|^{\frac{3}{2}} + |B|^{\frac{3}{2}} \right) + \gamma(k+2)^{\frac{3}{2}} n^{\frac{3}{2}}$$

Notice that, since $n > \beta(k+2)^3$, we have that $|X| \leq (1 - \sqrt{\frac{2}{3}})n$. Since $|A|, |B| \leq \frac{2}{3}n$, we find:

$$
\begin{aligned}
|A|^{\frac{3}{2}} + |B|^{\frac{3}{2}} &\leq \left(\frac{2}{3}n \right)^{\frac{3}{2}} + \left(\frac{1}{3}n - |X| \right)^{\frac{3}{2}} \\
&\leq \left(\frac{2}{3}n \right)^{\frac{3}{2}} + \left(n\sqrt{\frac{2}{3}} - \frac{2}{3}n \right)^{\frac{3}{2}} \leq \left(\frac{2}{3} \right)^{\frac{3}{4}} n^{\frac{3}{2}}
\end{aligned}
$$

By chosing $C \geq \gamma/(1 - (2/3)^{\frac{3}{4}})$, the result follows. $\square$

Notice that this also proves that if $G = (V, E)$ is a graph with n vertices and of treewidth k, then the pathwidth is at most $O(k^{\frac{3}{2}} \sqrt{n})$. Lemma 2.2.10 on page 23 shows that in fact a better upper bound holds; the pathwidth of a graph is at most $\log n$ times the treewidth of a graph.

6.2 Approximating treewidth

In this section we present a polynomial time algorithm to find a $\log n$ approximation for the treewidth of a graph. As reported in [95], Leighton et al. [118] obtain the following theorem.

> There exists a polynomial time algorithm that, given a graph $G = (V, E)$ and a set $W \subseteq V$ of vertices, finds a $\frac{2}{3}$-vertex-separator W in G of size $O(k \log n)$, where k is the minimum size of a $\frac{1}{2}$-vertex-separator of W in G.

Recall that if G is a graph with treewidth $\leq k$ and W is a subset of vertices, then there is a $\frac{1}{2}$-vertex-separator of W in G with size at most $k+1$. We conclude the following:

Theorem 6.2.1 *There exists a polynomial time algorithm and a constant Δ that, given a graph $G = (V, E)$ and a set $W \subseteq V$, finds a $\frac{2}{3}$-vertex-separator S of W in G of size $|S| \leq \Delta k \log(n+1)$, where $n = |V|$ and k is the treewidth of G.*

In the remainder of this section, Δ is assumed to be the constant mentioned in this theorem.

Definition 6.2.1 *Let $G = (V, E)$ be a graph. Let $W \subseteq V$. A W-partition is a partition $(X, (S_i)_{i=1,\dots,4})$ of the vertices in five sets, such that:*

1. *for $i \neq j$, X separates S_i and S_j,*
2. *for $i = 1, \dots, 4$: $|S_i| \leq \frac{2}{3}n$, and*
3. *for $i = 1, \dots, 4$: $|S_i \cap W| \leq \frac{2}{3}|W|$.*

Lemma 6.2.1 *There is a polynomial time algorithm $\mathcal{A}$, which takes as input an integer k, a graph $G = (V, E)$ (with $|V| = n$), and a subset of the vertices W and that outputs either:*

- *a correct statement that the treewidth of G is larger than k, or*

- *a W-partition $(S, (S_i)_{i=1,\dots,4})$ with $|S| \leq 3\Delta k \log(n+1)$.*

Proof. If the treewidth of G is not larger than k then the algorithm of Theorem 6.2.1 finds a $\frac{2}{3}$ vertex separator X of V with $|X| \leq \Delta k \log(n+1)$. If this separator is not found, then the treewidth of G exceeds k. Otherwise, find a partition (X, A, B) of V in G (see the proof of Lemma 6.1.2). Take a $\frac{2}{3}$ vertex separator U_1 of W in $G[X \cup A]$ and a $\frac{2}{3}$ vertex separator U_2 of W in $G[X \cup B]$ using Theorem 6.2.1. If $|U_1|$ or $|U_2|$ is larger than $\Delta k \log n$ then the treewidth of G exceeds k. Otherwise, in the same manner as described in the proof of Lemma 6.1.2, we group the the components of $G[(X \cup A) \setminus U_1]$ in two sets T_1 and T_2, such that $|T_i \cap W| \leq \frac{2}{3}|W|$. Similar, we group the components of $G[(X \cup B) \setminus U_2]$ in two sets T_3 and T_4. Let $S = X \cup U_1 \cup U_2$ and let $S_i = T_i \setminus X$ ($i = 1, \dots, 4$). Clearly $(X, (S_i)_{i=1,\dots,4})$ is a W-partition with $|S| \leq 3\Delta k \log(n+1)$. $\square$

In figure 6.2 (page 68) we describe an algorithm that, for a given graph $G = (V, E)$ and a given subset of the vertices W, triangulates G such that W is a clique and such that the clique number is at most $\max(\frac{4}{3}|W|, 12k\Delta \log(n+1))$, or outputs that the treewidth of this graph is larger than k. Hence for $W = \emptyset$ the algorithm triangulates G such that the clique number is not larger than $12k\Delta \log(n+1)$ or outputs that the treewidth is larger than k.

procedure $\text{triang}(G, W, k)$;

comment {Input: Graph $G = (V, E)$ *with* $|V| = n$,
       subset of vertices W *and integer* k.
       Output: A triangulation of G *such that* W *is a clique and*
       such that $\omega \leq \max(\frac{4}{3}|W|, 12k\Delta \log(n+1))$ *or,*
       a valid statement that the treewidth of G *is larger than* k}.

begin
  $\beta \leftarrow 12k\Delta$;
  if $n \leq \beta \log(n+1)$ **then**
    make a clique of G
  else
    begin
      find a W-partition $(X, (S_i)_{i=1,...,4})$ with $|X| \leq 3k\Delta \log(n+1)$ in G;
      if such a W-partition is not found **then**
              ouput("the treewidth exceeds k");
    else
      begin
        for $i \leftarrow 1$ **to** 4 **do**
            begin
               $W_i \leftarrow S_i \cap W$;
               call $\text{triang}(G[S_i \cup X], W_i \cup X, k)$
            end;
        add edges between vertices of $W \cup X$, making a clique of $G[W \cup X]$
      end
    end
end

Figure 6.2: Procedure to triangulate G such that W is a clique.

Lemma 6.2.2 *If* $G = (V, E)$ *is a graph with* n *vertices,* k *an integer and* W *a subset of* V, *then* $\mathrm{triang}(G, W, k)$ *triangulates* G *such that the vertices of* W *form a clique and such that the clique number is at most*

$$\max\left(\frac{4}{3}|W|, 12k\Delta \log(n+1)\right)$$

or correctly outputs that the treewidth of G *is larger than* k.

Proof. If the algorithm outputs that the treewidth of G is larger than k, then this is a correct statement by Lemma 6.2.1. Assume the algorithm does not output that the treewidth exceeds k. First we show that the algorithm terminates. Let $\beta = 12k\Delta$. If $n > \beta \log(n+1)$, then $|X \cup S_i| \le 3k\Delta \log(n+1) + \frac{2}{3}n < \frac{11}{12}n$. This guarantees termination.

Now we show that the algorithm returns a triangulated graph. We prove this by induction using the recursive structure of the algorithm. Clearly the claim is true if $n \le \beta \log(n+1)$. Assume $n > \beta \log(n+1)$. By induction the recursive calls $\mathrm{triang}(G[S_i \cup X], W_i \cup X, k)$ return triangulations of $G[S_i \cup X]$, such that $W_i \cup X$ is a clique. Finally the algorithm makes a clique of $W \cup X$. Notice that, for $i = 1, \ldots, 4$, the graphs $G[S_i \cup W \cup X]$ are triangulated. Since the intersection of these triangulated graphs is a clique, the union is triangulated.

We show that the clique number of this triangulation is at most $\max(\frac{4}{3}|W|, \beta \log(n+1))$. This is clearly true if $n \le \beta \log(n+1)$. Hence assume $n > \beta \log(n+1)$. Let M be a maximal clique. If M contains no vertex of $S_i \setminus W_i$, for $i = 1, \ldots, 4$, then M contains only vertices of $W \cup X$. By Lemma 6.2.1 we have $|W \cup X| \le |W| + 3k\Delta \log(n+1)$. If $|W| \le (\beta - 3k\Delta) \log(n+1)$ then $|W \cup X| \le \beta \log(n+1)$. Otherwise, $|W \cup X| \le \frac{4}{3}|W|$.

Now assume M contains a vertex of $S_i \setminus W_i$. Then clearly, it contains no vertex of S_j for $j \ne i$. Hence M is a clique in the triangulation of $G[S_i \cup X]$. By induction we know

$$|M| \le \max\left(\frac{4}{3}|W_i \cup X|, \beta \log(1 + |S_i \cup X|)\right)$$

Notice that:

$$\beta \log(1 + |S_i \cup X|) \le \beta \log(n+1)$$

and:

$$\frac{4}{3}|W_i \cup X| \le \frac{8}{9}|W| + 4k\Delta \log(n+1).$$

If $\frac{8}{9}|W| \le (\beta - 4k\Delta) \log(n+1)$ then $\frac{4}{3}|W_i \cup X| \le \beta \log(n+1)$, and otherwise $\frac{4}{3}|W_i \cup X| \le (\frac{8}{9} + \frac{4}{9})|W| = \frac{4}{3}|W|$. Hence $|M| \le \max(\frac{4}{3}|W|, \beta \log(n+1))$. This proves the lemma. $\qquad\square$

Theorem 6.2.2 *There exists a polynomial time algorithm B which takes as input an integer k and a graph $G = (V, E)$ with n vertices and that outputs either:*

- *a correct statement that the treewidth of G exceeds k, or*

- *a triangulation of G with clique number at most $12k\Delta \log(n+1)$.*

If algorithm A can be implemented to run in time $O((nk)^c)$ for some constant $c > 1$, then also B can be implemented to run in time $O((nk)^c)$.

Proof. The algorithm B consists of calling $\text{triang}(G, \emptyset, k)$. By Lemma 6.2.2 the algorithm either outputs that the treewidth exceeds k, or triangulates G with a clique number not larger than $12k\Delta \log(n+1)$. Hence we only have to bound the running time. Assume that A runs in $O((nk)^c)$ for some $c > 1$.

Let $T_k(n, w)$ be the running time for $\text{triang}(G, W, k)$, if G is a graph with n vertices and W a set with w vertices. We show there exist constants α and β such that:

$$T_k(n, w) \le \alpha(nk)^c + \beta nw$$

Taking $w = 0$ proves the theorem. Let A be some constant larger than 12 (we fix A later). We start by showing that if $n \le Ak\Delta \log(n+1)$ then the formula holds. Consider the *recursion tree*, which is defined recursively as follows. If no recursive call is made, the recursion tree consists of one node. Otherwise, take a new root node corresponding to the calling program, and make this adjacent to the roots of the recursion trees corresponding to the recursive calls. As we have shown before (in the proof of Lemma 6.2.2), each recursive call decreases the number of vertices by at least a factor $\frac{11}{12}$. We show that the fraction $n/(k \log(n+1))$ decreases by at least a constant factor, thus showing that the depth of the recursion tree is bounded by some constant. Clearly if $n \le 8$ the depth of the recursion tree is bounded by a constant. Hence we may assume $\ln(n+1) \ge 2$.

$$\frac{11n/12}{k \log(11n/12 + 1)} \cdot \frac{k \log(n+1)}{n} \le \frac{11}{12} \cdot \frac{\log(n+1)}{\log(11/12) + \log(n + 12/11)}$$

$$\le \frac{11}{12} \cdot \frac{\log(n+1)}{\log\frac{11}{12} + \log(n+1)} \le \frac{11}{12} \cdot \frac{1}{1 - \frac{\log(12/11)}{\log(n+1)}}$$

$$\le \frac{11}{12} \cdot \frac{1}{1 - \frac{1}{2}\ln(12/11)} \le \frac{11}{12} \cdot \frac{1}{1 - \frac{1}{22}} < 1.$$

This shows that the depth of the recursion tree is bounded by some constant D. Each node in the recursion tree has at most four children, hence the recursion tree contains at most a number of nodes proportional to 4^D. We show that the extra amount of time needed for each node is $O((nk)^c)$.

Finding the W-partition takes $O((nk)^c)$ time. Apart from the recursive calls and from finding the W-partition, the time needed is $O(n^2)$. If $n < k$ then $O(n^2) = O((nk)^c)$, and otherwise since $n \leq Ak\Delta \log(n+1)$, there is a constant C such that:

$$n^{1/c} \leq \frac{Cn}{\log(n+1)} \leq ACk\Delta \Rightarrow n^2 \leq (AC\Delta)^{2c}k^{2c} \leq (AC\Delta)^{2c}(nk)^c.$$

This shows that the running time is $O((nk)^c)$ in case $n \leq Ak\Delta \log(n+1)$.

Let $n > 12k\Delta \log(n+1)$. We show a recurrence relation for $T(n,w)$. We assumed that finding the W-partition can be done in $O((nk)^c)$ time. The only other non-trivial part is the construction of the clique $W \cup X$, after the recursive calls are made. The recursive calls $\mathrm{triang}(G[S_i \cup X], W_i \cup X, k)$ make cliques of $W_i \cup X$. Hence, when making a clique of $W \cup X$, we only have to introduce edges between vertices x and y which are in different sets W_i and W_j. But these vertices were not adjacent before. This shows that introducing such an edge can be done in $O(1)$ time. The number of edges in the triangulated graph is at most $n \cdot \max(\frac{4}{3}|W|, 12k\Delta \log(n+1))$. This shows there exists constants γ and δ such that:

$$T(n,w) \leq \sum_{i=1}^{4} T(n_i + |X|, w_i + |X|) + \gamma(kn)^c + \delta nw$$

where $n_i = |S_i|$ and $w_i = |W_i|$. By induction:

$$T(n,w) \leq \sum_{i=1}^{4} \{\alpha k^c(n_i + |X|)^c + \beta(n_i + |X|)(w_i + |X|)\} + \gamma(kn)^c + \delta nw$$

Notice that since $w_i \leq \frac{2}{3}w$, $\sum_{i=1}^{4} n_i = n - |X|$ and $|X| \leq 3k\Delta \log(n+1)$:

$$\sum_{i=1}^{4} \beta(n_i + |X|)(w_i + |X|) \leq \beta \left(\frac{2}{3}w + 3k\Delta \log(n+1)\right)(n + 9k\Delta \log(n+1))$$

We can choose A large enough such that $9k\Delta \log(n+1) \leq \frac{1}{5}n$ for all $n > Ak\Delta \log(n+1)$. Then, if $\beta \geq 5\delta$:

$$\frac{2}{3}\beta w(n + 9k\Delta \log(n+1)) + \delta nw \leq \frac{4}{5}\beta nw + \delta nw \leq \beta nw$$

Clearly, we can choose a new constant γ' such that:

$$\beta \cdot 3k \log(n+1)(n + 9k\Delta \log(n+1)) + \gamma(kn)^c \leq \gamma'(nk)^c$$

Hence we obtain, for $n > Ak\Delta \log(n+1)$:

$$T(n,w) \leq \alpha k^c \sum_{i=1}^{4} (n_i + |X|)^c + \beta nw + \gamma'(nk)^c$$

Without loss of generality, we may assume that $n_1 \geq \frac{1}{4}(n - |X|)$. Then it is easily checked that:

$$
\begin{aligned}
\sum_{i=1}^{4}(n_i + |X|)^c &\leq (n_1 + |X|)^c + \left(\sum_{i=2}^{4}(n_i + |X|)\right)^c \\
&= (n_1 + |X|)^c + (n - n_1 + 2|X|)^c \\
&\leq \left(\frac{1}{4}n + \frac{3}{4}|X|\right)^c + \left(\frac{3}{4}n + \frac{9}{4}|X|\right)^c \\
&\leq \left(\left(\frac{1}{4}\right)^c + \left(\frac{3}{4}\right)^c\right)\left(1 + 9k\Delta\frac{\log(n+1)}{n}\right)^c n^c
\end{aligned}
$$

It follows that there is a constant $\kappa < 1$ such that if we choose A large enough then for all $n > Ak\Delta \log(n+1)$: $\sum_{i=1}^{4}(n_i + |X|)^c \leq \kappa n^c$.

If we choose $\alpha \geq \frac{\gamma'}{1-\kappa}$ then:

$$T(n, w) \leq \kappa\alpha(nk)^c + \beta nw + \gamma'(nk)^c \leq \alpha(nk)^c + \beta nw$$

which proves the theorem. $\square$

We end this section by showing an approximation algorithm for pathwidth with performance ratio $O(\log^2 n)$. Recall Lemma 2.2.10. It is easily seen that the proof of this lemma can be made constructive in case the graph is triangulated:

> There is a polynomial time algorithm that given a triangulated graph H with n vertices and treewidth k, finds a path-decomposition for H with width at most $(k+1)\log(n) - 1$.

This proves the following theorem.

Theorem 6.2.3 *There exists a polynomial algorithm that, given a graph G with n vertices and with treewidth k, finds a path-decomposition of G with pathwidth $O(k\log^2 n)$.*

6.3 Absolute approximations

In this section we show that if $P \neq NP$, then no absolute approximation algorithms exist for treewidth and for pathwidth.

Given an approximation algorithm $\mathcal{A}$ for a minimization problem, we can distinguish between three kinds of performance guarantees. First, in an *absolute approximation*, the approximate solution $\mathcal{A}(I)$ is within an additive constant of the optimal solution $\mathcal{OPT}(I)$. Second, the approximate solution can be within a multiplicative constant of the optimal one. Finally, the ratio between the optimal and approximate solutions can grow with the size of the problem. The algorithms we have presented above all have performance guarantees of the latter kind, with ratios of $O(\sqrt{kn})$, $O(\log n)$ or $O(\log^2 n)$. The hardest of these bounds to achieve is the absolute bound; few NP-hard problems have absolute approximation algorithms.

Recall Lemma 2.2.2 (page 19). Given a tree-decomposition (S, T) for a graph $G = (V, E)$, and let $W \subseteq V$ induce a clique in G. Then there is a subset $X \in S$, such that $W \subseteq X$.

Theorem 6.3.1 *If* $P \neq NP$*, then no polynomial time approximation algorithm* $\mathcal{A}$ *for the treewidth problem can guarantee* $\mathcal{A}(G) - \mathcal{OPT}(G) \leq C$ *for a fixed constant* C.

Proof. Assume we have a polynomial time algorithm $\mathcal{A}$ that, given a graph $G = (V, E)$, finds a tree-decomposition of G with treewidth at most C larger than the treewidth of G. Let a graph $G = (V, E)$ be given. Let $G' = (V', E')$ be the graph obtained by replacing every vertex of G by a clique of $C + 1$ vertices, and adding edges between every pair corresponding to adjacent vertices in G. Thus

$$V' = \{v_i \mid v \in V, 1 \leq i \leq C + 1\} \text{ and}$$
$$E' = \{(v_i, w_j) \mid (v = w \wedge i \neq j) \vee (v, w) \in E\}$$

We examine the relationship between the treewidth of G and the treewidth of G'.

Suppose we have a tree-decomposition (X, T), with $X = \{X_i \mid i \in I\}$ and $T = (I, F)$ of G with treewidth L. One easily checks that

$$(\{Y_i \mid i \in I\}, T = (I, F)) \text{ with } Y_i = \{v_j \mid v \in X_i, 1 \leq j \leq C + 1\}$$

is a tree-decomposition of G' with treewidth $(L + 1)(C + 1) - 1$. It follows that:

$$\text{treewidth}(G') \leq (\text{treewidth}(G) + 1) \cdot (C + 1) - 1$$

Next suppose we have a tree-decomposition (Y, T) with $Y = \{Y_i \mid i \in I\}$ and $T = (I, F)$ of G' with treewidth M. Let $X_i = \{v \in V \mid \{v_1, v_2, \cdots, v_{C+1}\} \subseteq Y_i\}$. We claim that (X, T) with $X = \{X_i \mid i \in I\}$ is a tree decomposition of G with treewidth $(M + 1)/(C + 1) - 1$. Let $(v, w) \in E$. Note that $v_1, v_2, \cdots, v_{C+1}, w_1, w_2, \cdots, w_{C+1}$ form a clique in G'. Hence, by Lemma 2.2.2 there exists an $i \in I$ with $\{v_1, \cdots, v_{C+1}, w_1, \cdots w_{C+1}\} \subseteq Y_i$, and thus $v, w \in X_i$. Let $j \in I$ be on the path in T from $i \in I$ to $k \in I$. If $v \in X_i \cap X_k$, then $\{v_1, \cdots, v_{C+1}\} \subseteq Y_i \cap Y_k$, and hence by definition of a tree-decomposition $\{v_1, \cdots, v_{C+1}\} \subseteq Y_j$, so $v \in X_j$. Clearly, $\max_{i \in I} |X_i| \cdot (C + 1) \leq \max_{i \in I} |Y_i|$. This finishes the proof of our claim. It follows that $\text{treewidth}(G') \geq (\text{treewidth}(G) + 1) \cdot (C + 1) - 1$ and hence:

$$\text{treewidth}(G') = (\text{treewidth}(G) + 1) \cdot (C + 1) - 1$$

Now consider the following algorithm for the treewidth problem. Let G be the input graph. Form G', and apply algorithm $\mathcal{A}$ to G'. Apply the construction described above to form a tree-decomposition of G. This must be a tree-decomposition with minimum treewidth: if the treewidth of G is k, then the treewidth of G' is $(k+1)(C+1)-1$. Hence, $\mathcal{A}$ outputs a tree-decomposition of G' with treewidth at most $(k+1)(C+1)-1+C$, and the algorithm described above outputs a tree-decomposition of G with treewidth at most $\lfloor ((k+1)(C+1)+C)/(C+1)-1 \rfloor = k$. Thus we would have a polynomial time algorithm for treewidth contradicting the NP-completeness of the problem.   □

In the same way we can prove the following theorem. This result was also proved (with different terminology) by Deo et al. [54].

Theorem 6.3.2 *If* $P \neq NP$, *then no polynomial time approximation algorithm* $\mathcal{A}$ *for the pathwidth problem can guarantee* $\mathcal{A}(G) - \mathcal{OPT}(G) \leq C$ *for a fixed constant* C.

Chapter 7

Approximating treewidth and pathwidth for some classes of perfect graphs

For some *special classes* of graphs, it has been shown that the treewidth can be computed efficiently. In this chapter we discuss the problem of finding approximate tree- and path-decompositions for cotriangulated graphs, convex graphs, permutation graphs and for cocomparability graphs. We also show that for these graphs, if the treewidth is at most k, then the pathwidth is bounded by some polynomial in k. Our results show that it is oftern very easy to find good approximations for tree- and pathwidth.

It has been shown that computing the pathwidth is NP-hard [94], even when restricted to triangulated graphs [78]. It is unknown whether there exist good approximations, which can be computed efficiently, for the pathwidth of a triangulated graph. Surprisingly, for *cotriangulated graphs* (complements of triangulated graphs), the pathwidth and treewidth are related in a linear fashion, and there is a very simple algorithm that computes an approximate path-decomposition. We show this in section 7.2. Perhaps even more surprising, the exact pathwidth and treewidth of cotriangulated graphs can be computed in polynomial time. We show this in subsection 7.2.1.

It is easy to see that computing the treewidth and pathwidth for bipartite graphs is NP-complete. Indeed, finding an approximation algorithm with a certain performance ratio for bipartite graphs is at least as difficult as finding an approximation with the same performance for graphs in general. This can be seen as follows. We need the concept called a subdivision graph (see [82, pages 79–80]).

Definition 7.0.1 *Given a graph* G, *the* subdivision graph, $S(G)$ *is obtained by placing a new vertex on each edge of the graph, and by replacing the edge by two edges incident with the new vertex.*

Clearly, $S(G)$ is bipartite. The treewidth of $S(G)$ is equal to the treewidth of G. Given a tree-decomposition for $S(G)$ it can be easily transformed into a tree-decomposition for G with the same width. This shows that finding an approximation for the treewidth of bipartite graphs is as hard as finding approximations for the treewidth in general.

In section 7.3 we show that, if the bipartite graph is *convex*, with treewidth k, the pathwidth is at most $2k + 1$ and it is almost straightforward to find a path-decomposition with width $2k + 1$ in $O(nk)$ time. In chapter 8, we show that computing the treewidth of chordal bipartite graphs (or graphs that are weakly chordal and bipartite) is polynomial. Chordal bipartite graphs are bipartite graphs such that every induced cyle of length at least six has a chord. There is a strong connection between chordal bipartite graphs, strongly chordal graphs and totally balanced matrices. Indeed, a graph is chordal bipartite if and only if the adjacency matrix is totally balanced [60]. As the class of convex graphs is properly contained in the class of chordal bipartite graphs, this shows that the exact treewidth of convex graphs can be computed in polynomial time. However, the running time of this algorithm is not very good ($O(m^\alpha)$, i.e., the time needed to compute the product of two $m \times m$ matrices where m is the number of edges in the graph), and we think that, especially for practical applications, the algorithm described in this chapter, is of importance. Furthermore, as far as we know, computing the exact pathwidth for chordal bipartite graphs is still an open problem.

One of the most well-known and well-studied classes of perfect graphs is the class of *permutation graphs*. Permutation graphs are exactly the comparability graphs of posets of dimension at most two (see below). They can also be characterized as the graphs which are both comparability and cocomparability graphs. We show in section 7.4 that if the treewidth of a permutation graph is k, then the pathwidth is at most $2k$ and there is a linear time algorithm which produces a path-decomposition with this width. In chapter 9 we show that the pathwidth and treewidth of a permutation graph are equal, and we show a linear time algorithm computing the exact treewidth of a permutation graph. We included the approximation algorithm of section 7.4 for two reasons. First, this algorithm can be used as an intermediate step in the exact algorithm and, second, it illustrates nicely the method used for approximating the pathwidth of cocomparability graphs in general.

Comparability graphs and their complements have a large number of applications (see e.g. [76, 141]). These graph classes, as well as those of the triangulated and of the cotriangulated graphs, are among the largest and most important classes of perfect graphs. Cocomparability graphs properly contain many other well known classes of perfect graphs, like interval graphs, cographs (or P_4-free graphs), permutation graphs, indifference graphs, complements of superperfect graphs etc. (see e.g. [76] or [15, pages 67–96]). Cocomparability graphs are exactly the intersection graphs of the graphs of continuous functions $F_i : (0, 1) \rightarrow \mathbf{R}$. See [77] for a proof of this. Comparability graphs can be recognized in polynomial time [135, 76, 72, 155]. Given a comparability graph, a transitive orientation of the edges can be found in $O(n^2)$ time [155].

We show in section 7.5 that, if the treewidth of a cocomparability graph G is at most k, then the pathwidth is at most $O(k^2)$, and we give a simple algorithm which finds a path-decomposition with this width. In [30] we show that in fact the treewidth and pathwidth of a cocomparability graph are equal. However, computing the exact treewidth of a cocomparability graph is NP-complete. This follows from the results in [4]. As far as we know, ours is the first algorithm for approximating the treewidth of cocomparability graphs. We do not know if an approximation for treewidth with a constant performance ratio is possible. Our algorithm uses a transitive orientation of the complement $\overline{G}$, and the *height function* which can be obtained from this orientation in linear time. The algorithm also uses a vertex coloring of the graph G. This vertex coloring, or clique cover in the complement, can be found in $O(n^3)$ time using a flow algorithm, as described in [76].

7.1 Preliminaries

In this section we start with some definitions and easy lemmas. For more information on special perfect graph classes the reader is referred to [38, 76]. An examaple of a class of perfect graphs is the class of convex graphs (in fact all bipartite graphs are perfect).

Definition 7.1.1 *Let* $G = (X, Y, E)$ *be a bipartite graph. An ordering of* X *has the* adjacency property, *if for each* $y \in Y$*, the neighbors of* y *in* X *are consecutive in the ordering of* X*.*

Definition 7.1.2 *A bipartite graph* $G = (X, Y, E)$ *is* convex, *if there is an ordering of* X *or of* Y *with the adjacency property.*

A bipartite graph $G = (X, Y, E)$ is *biconvex* if there is an ordering of X *and* Y with the adjacency property. Convex graphs contain the bipartite permutation graphs. Notice that convex graphs can be recognized in linear time, using for example the PQ-tree algorithms of Booth and Lueker [35] (see also [38, 154]). Recall the Definition 3.1.1, on page 28, of a comparability graph. We know that if a graph G is a comparability graph, then this holds for every induced subgraph of G. There exists a complete list of critical noncomparability graphs [68]. Comparability graphs can be recognized in time $O(n^\alpha)$, which is the time needed to square the 0/1-adjacency matrix [155].

By the perfect graph theorem, also cocomparability graphs, which are the complements of comparability graphs (Definiton 3.1.1, page 28), are perfect. For our algorithm we shall need the concept of a height function defined in [76]. Let F be an acyclic orientation of an undirected graph $G = (V, E)$. A *height function* h assigns a non-negative integer to each vertex as follows: $h(v) = 0$ if v is a sink; otherwise, $h(v) = 1 + \max\{h(w) \mid (v, w) \in F\}$. In other words, $h(v)$ is the maximal length of a path from v to a sink. A height function can be assigned in linear time [76], and represents a proper vertex coloring of G. If F is a transitive orientation, the coloring by h is optimal (i.e. uses the least possible number of colors) [76].

We also need a coloring of the cocomparability graph. In [76] a method is described to find an optimal coloring of a cocomparability graph by using a minimum flow algorithm. It follows that this coloring can be found in $O(n^3)$ time. Recently this result was improved upon in [154]. Here an $O(\sqrt{n}m)$ algorithm is given to find a clique partition of a comparability graph.

Cocomparability graphs are intersection graphs [77].

Lemma 7.1.1 *A graph* G *with* n *vertices is a cocomparability graph if and only if* G *is the intersection graph of the graphs of* n *continuous functions* $F_i : (0, 1) \to \mathbf{R}$.

We only use this lemma implicitly.

We think of a permutation π of the numbers $1, \dots, n$, as the sequence $\pi = [\pi_1, \dots, \pi_n]$. We use the notation π_i^{-1} for the position of the number i in this sequence.

Definition 7.1.3 *If* π *is a permutation of the numbers* $1, \dots, n$, *we can construct an undirected graph* $G[\pi] = (V, E)$ *with vertex set* $V = \{1, \dots, n\}$, *and edge set* E:

$$(i, j) \in E \Leftrightarrow (i - j)(\pi_i^{-1} - \pi_j^{-1}) < 0$$

An undirected graph is called a permutation graph *if there exists a permutation* π *such that* $G \cong G[\pi]$.

$G[\pi]$ is sometimes called the *inversion graph* of π. Notice that we can obtain the complement of $G[\pi]$, by reversing the sequence π. Hence the complement of a permutation graph is also a permutation graph. It is also easy to see that a permutation graph is a comparability graph. Pnueli, Lempel and Even ([135]) showed that a graph G is a permutation graph if and only if both G and $\overline{G}$ are comparability graphs. It follows that permutation graphs are *perfect*. They can be recognized in time $O(n^2)$ (see [155]). There exist fast algorithms for many NP-complete problems like CLIQUE, INDEPENDENT SET, FEEDBACK VERTEX SET and DOMINATING SET when restricted to permutation graphs [76, 61, 40, 39].

In this chapter we assume that the permutation π is given, and we show some results on the pathwidth and treewidth of $G[\pi]$, which is sometimes called the *inversion graph* of π. If the permutation π is *not* given, transitive orientations of G and $\overline{G}$ can be computed in $O(n^2)$ time [155]. Given these orientations, a permutation can be computed in $O(n^2)$ time [76].

Every permutation graph $G[\pi]$ is an intersection graph, which is illustrated by the *matching diagram* of π [76], see figure 7.1.

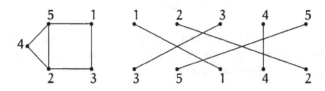

Figure 7.1: permutation graph and matching diagram

Definition 7.1.4 *Let π be a permutation of $1, \ldots, n$. The matching diagram of π can be obtained as follows. Write the number $1, \ldots, n$ horizontally from left to right. Underneath, write the numbers $\pi_1, \ldots, \pi_n$, also horizontally from left to right. Draw n straight line segments joining the two 1's, the two 2's, etc.*

Notice that two vertices i and j of $G[\pi]$ are adjacent if and only if the corresponding line segments in the matching diagram of π intersect. Matching diagrams are often useful in visualizing certain concepts.

7.2 Splitgraphs and cotriangulated graphs

Recall that for triangulated graphs the treewidth is equal to the clique number minus one. It follows that, for triangulated graphs, the treewidth can be computed in linear time. Computing the pathwidth of a triangulated graph is an NP-complete problem, as shown in [78].

We show in this section that the treewidth and pathwidth of cotriangulated graphs can be computed efficiently. We start by showing how to find good approximations for the treewidth and pathwidth of cotriangulated graphs. Let G be a cotriangulated graph with treewidth at most k. We show that the pathwidth of G is at most $3k + 4$ and there exists an $O(n^2)$ algorithm that finds a path-decomposition with this width. According to Lemma 2.1.4, page 9, if $H = (V, E)$ is a triangulated graph with n vertices then there is a clique C such that every component of $H[V - C]$ has at most $\lceil \frac{1}{2}(n - |C|) \rceil$ vertices. Also Lemma 2.2.1, page 18, will be useful: for the complete bipartite graph $K(m, n)$, the treewidth is $\min(m, n)$.

Let C be a clique in $\overline{G}$ as mentioned in the lemma 2.1.4. The vertices of $V \setminus C$ can be partitioned in two sets A and B such that no vertex of A is adjacent to a vertex of B, and both A and B have at most $\lceil \frac{2}{3}(n - |C|) \rceil$ vertices. Notice that the subgraph induced by A and B has a complete bipartite subgraph in G, since every vertex of A is adjacent to every vertex of B. As the treewidth of G is at most k and by lemma 2.2.1:

$$\lfloor \frac{1}{3}(n - |C|) \rfloor \leq \min(|A|, |B|) \leq k$$

Hence $|A \cup B| \leq 3(k + 1)$. We can triangulate G by adding edges such that $A \cup B$ becomes a clique. Since C is a stable set in G, the result is a splitgraph. For a splitgraph we have the following result.

Lemma 7.2.1 *Let H be a splitgraph with maximum size clique C. Then the treewidth of H is $|C| - 1$. The pathwidth is either $|C| - 1$ or $|C|$. The pathwidth of H is equal to the treewidth if and only if there are vertices x and y in C (possibly equal) such that $N(x) \cap N(y) \subseteq C$.*

Proof. Since H is triangulated the treewidth is equal to the clique number minus one. We can construct a path-decomposition of width $|C|$ as follows. Let $y_1, y_2, \ldots, y_s$ be the vertices of $V \setminus C$. Construct a path-decomposition $(X_1, X_2, \ldots, X_s)$ with $X_i = C \cup \{y_i\}$ for $i = 1, \ldots, s$. It is easy to check that this is indeed a path-decomposition.

Assume there exists a path-decomposition $P = (X_1, \ldots, X_t)$ of width $|C| - 1$. By Lemma 2.2.2, page 19, there exists a subset X_i such that $C \subseteq X_i$. Since the width of P is $|C| - 1$, we have $X_i = C$. We can assume $X_{i+1} \neq X_i$ and $X_{i-1} \neq X_i$ (take $X_0 = \emptyset$ and $X_{t+1} = \emptyset$). Let $x \in X_i \setminus X_{i+1}$ and $y \in X_i \setminus X_{i-1}$. Let $z \in N(x) \cap N(y)$. Then z must be an element of a subset X_p which also contains x and of a subset X_q which also contains y. Since $p \leq i$ and $q \geq i$ we must have $z \in X_i$. This shows that $N(x) \cap N(y) \subseteq C$.

Finally assume there are vertices x and y in C with $N(x) \cap N(y) \subseteq C$. Let $z_1, \ldots, z_r$ be the vertices of $N(x) \setminus C$ and let $z_{r+2}, \ldots, z_t$ be the vertices of $V \setminus (C \cup \{z_1, \ldots, z_r\})$. Define subsets X_i for $i = 1, \ldots, t$ as follows:

$$X_i = \begin{cases} \{z_i\} \cup (C \setminus \{y\}) & \text{for } i = 1, \ldots, r \\ C & \text{if } i = r + 1 \\ \{z_i\} \cup (C \setminus \{x\}) & \text{for } i = r + 2, \ldots, t \end{cases}$$

It is easy to check that this gives a correct path-decomposition of width $|C| - 1$. $\qquad \square$

As an immediate consequence we have the following result.

Theorem 7.2.1 *If* G *is a cotriangulated graph with treewidth at most* k, *then the pathwidth is at most* $3k+4$ *and there exists an* $O(n^2)$ *algorithm which produces a path-decomposition with this width.*

In the next subsection we show how to find the exact treewidth and pathwidth of a cotriangulated graph. The following result will be useful.

Lemma 7.2.2 *Let* H *be a splitgraph. There is an algorithm to find a path-decomposition of width at most* $\omega(H)$. *This algorithm can be implemented to run in time* $O(\omega(H)n)$ *(where* n *is the number of vertices of* H*). The exact pathwidth of* H *can be found in time* $O(n^\alpha)$, *which is the time needed to square the* 0/1*-adjacency matrix of* H *(currently* $\alpha = 2.37...$*).*

Proof. Since H is triangulated, a maximum clique C can be found in $O(|V| + |E|)$ time [76]. As is shown in the proof of Lemma 7.2.1, a path-decomposition of width at most $|C|$ can be found in time $O(\omega(H)n)$. Using fast matrix multiplication techniques [117] and using the characterization of Lemma 7.2.1, the exact pathwidth can be found in $O(n^\alpha)$. □

Remark. In [78] an $O(n^3)$ algorithm is given to determine the pathwidth of a splitgraph. In fact a somewhat more general class of triangulated graphs is treated there.

7.2.1 Exact algorithms for the treewidth and pathwidth of cotriangulated graphs

In this subsection we give an algorithm to compute the treewidth of a co-triangulated graph. We start by showing that there exists a special tre-decomposition for triangulated graphs.

Definition 7.2.1 *Let* $G = (V, E)$ *be a triangulated graph. A* clique-tree-decomposition *for* G *is a tree-decomposition* $D = (S, T)$ *with* $T = (I, F)$ *and* $S = \{X_i | i \in I\}$ *such that for every* $i \in I$, X_i *is a maximal clique in* G *and such that for every* $i, j \in I$, *if* $i \neq j$ *then* $X_i \neq X_j$.

Lemma 7.2.3 *Let* $G = (V, E)$ *be triangulated. There is a* clique-tree-decomposition *for* G.

Proof. By induction. If G is a clique, then make a tree T with one node and a corresponding subset containing all vertices of G. Otherwise, let x be a simplicial vertex of G, and let C be the set of neighbors of x. By induction there is a clique-tree-decomposition $D' = (S', T')$ for $G[V \setminus \{x\}]$. Suppose there is a subset Y_i in S' with $Y_i = C$. Add X to this subset. The new

tree-decomposition is a clique-tree-decomposition for G. Suppose there is
no subset in S' equal to C. There must exist a subset Y_i such that $C \subset Y_i$.
Take a new node i_0 and make it adjacent to i in T'. Make a corresponding
subset $Y_{i_0} = \{x\} \cup C$. It is easy to check that the new tree-decomposition is
a clique-tree-decomposition for G.   □

We use the following notation. Let $G = (V, E)$ be triangulated, and
let $D = (S, T)$ be a clique-tree-decomposition for G with $T = (I, F)$ and
$S = \{X_i | i \in I\}$. Let $(i, j) \in F$. We write $T^i_{(i,j)}$ for the maximal subtree
containing i obtained by removing the edge (i, j). Let $I^i_{(i,j)}$ be the set of nodes
of $T^i_{(i,j)}$. Also we write $V^i_{(i,j)}$ for the following subset of vertices:

$$V^i_{(i,j)} = \{x | x \in X_t \setminus X_j \wedge t \in I^i_{(i,j)}\}$$

Lemma 7.2.4 *Let $G = (V, E)$ be triangulated and let $D = (S, T)$ be a
clique-tree-decomposition. Consider a node i in T. For each connected
component C of $G[V \setminus X_i]$ there is exactly one node j adjacent to i such
that $C \subseteq V^i_{(i,j)}$.*

Proof. Since C is connected, the nodes of T containing vertices of C must
form a (connected) subtree of T. Since X_i contains no vertices of C, the
lemma follows.   □

Lemma 7.2.5 *Let $G = (V, E)$ be triangulated, and let H be a triangulation
of the complement $\overline{G}$. Let C be a clique in G. There is at most one
connected component of $G[V \setminus C]$ with vertex set Y such that $H[Y]$ is not
a clique.*

Proof. Clearly, if x and y are vertices in different connected components of
$G[V \setminus C]$ then x and y are adjacent in H since they are already adjacent
in $\overline{G}$. Assume there are two different connected components with vertex
sets C_i and C_j such that $H[C_i]$ and $H[C_j]$ are not cliques. Then there are
vertices p and q in C_i which are not adjacent in H and vertices r and s in
C_j which are not adjacent in H. But then H cannot be triangulated since
$H[\{p, q, r, s\}]$ is a cycle of length four.   □

Theorem 7.2.2 *Let $G = (V, E)$ be triangulated. Let H be a triangulation
of $\overline{G}$. There is a maximal clique with vertex set C in G such that
$H[V \setminus C]$ is a clique.*

Proof. We show that there is a maximal clique C in G such that the vertex
sets of all connected components of $G[V \setminus C]$ induce cliques in H. Assume
this is not the case. Then by Lemma 7.2.5, for each maximal clique C in G

there is exactly one connected component which is not a clique in H. Call $\mathcal{C}(C)$ the vertex set of this connected component.

Let $D = (S, T)$ be a clique-tree-decomposition for G with $T = (I, F)$ and $S = \{X_i | i \in I\}$. Make a digraph with vertex set I as follows. Consider a node $i \in I$. By Lemma 7.2.4 there exists exactly one neighbor j of i in T such that all vertices of $\mathcal{C}(X_i)$ are contained in $V^i_{(i,j)}$. Direct an arc from i to j in this case. By assumption every node i gets an outgoing arc in this way. Since there are only arcs between nodes that are adjacent in T and since T is a tree, there must be nodes i and j which are adjacent in T and such that there is an arc from i to j and an arc from j to i. Let $A = V^i_{(i,j)} \setminus X_i$ and let $B = V^j_{(i,j)} \setminus X_j$. Then $H[A \cup B]$ is a clique.

Let $X^*_i = X_i \setminus X_j$ and $X^*_j = X_j \setminus X_i$. We claim that if $x \in X^*_i$ and $y \in X^*_j$ then x and y are adjacent in H. This can be seen as follows. Assume x and y are adjacent in G. Then there must be a subset X_t of S containing both x and y. But if $t \in I^i_{(i,j)}$ then by definition of a tree-decomposition, y must also be in X_i which is a contradiction. Similarly, t cannot be in $I^j_{(i,j)}$ and the statement follows. We also claim that each vertex of X^*_i is adjacent to each vertex of B in H and that each vertex of X^*_j is adjacent to each vertex of A. This follows by a similar argument.

We know that $H[V^j_{(i,j)}]$ is not a clique, since there is an arc from i to j. Let x and y be non adjacent vertices in $H[V^j_{(i,j)}]$. Since B is a clique, either both x and y are elements of X^*_j or one is in B and the other is in X^*_j. In both cases all vertices of A and all vertices of X^*_i are common neighbors of x and y in H. Since H is triangulated a minimal separator for x and y in H is a clique and any minimal separator must contain all common neighbors. It follows that $H[A \cup X^*_i]$ is a clique and hence also $H[A \cup B \cup X^*_i]$ is a clique. This proves the theorem.   □

Theorem 7.2.3 *Let* $G = (V, E)$ *be triangulated. If there is a maximum clique C (with* $\omega(G)$ *vertices) such that all vertices x in C have degree larger than* $\omega(G) - 1$ *in G, then the treewidth of* $\overline{G}$ *is* $|V| - \omega(G) - 1$. *Otherwise, the treewidth of G is* $|V| - \omega(G)$.

Proof. Let H be a traingulation of $\overline{G}$. By Theorem 7.2.2 there is a maximaml clique C in G such that $H[V \setminus C]$ is a clique. Notice that $\mathcal{S}(C)$ is a subgraph of H. It follows that we can restrict ourselves to splitgraphs $\mathcal{S}(C)$ with C a maximal clique in G. The treewidth of $\mathcal{S}(C)$ is either $|V| - |C| - 1$ or $|V| - |C|$. It follows that we can restrict ourselves to maximum cliques in G (with $\omega(G)$ vertices).

Let C be a maximum clique in G. Then the treewidth of $\mathcal{S}(C)$ is $|V| - |C| - 1$ if every vertex x in C is adjacent to some vertex of $V \setminus C$ in G (in that case $V \setminus C$ is a maxiomum clique in $\mathcal{S}(C)$). Otherwise the treewidth of $\mathcal{S}(C)$ is $|V| - |C|$.

Hence we may conclude that the treewidth of $\overline{G}$ is $|V| - \omega(G) - 1$ if there is a maximum clique C such that all vertices of C are adjacent to some vertex of $V \setminus C$ in G. Otherwise the treewidth of $\overline{G}$ is $|V| - \omega(G)$. □

Theorem 7.2.4 *Let $\overline{G}$ be a cotriangulated graph. There is an $O(n^2)$ algorithm to find a tree-decomposition with width equal to the treewidth of $\overline{G}$ (where n is the number of vertices of $\overline{G}$).*

Proof. We can construct the complement G of $\overline{G}$ in $O(n^2)$ time. For each vertex compute the degree in G. In [76] it is shown that a list of all maximum cliques can be constructed in $O(n^2)$ time. By Theorem 7.2.3 we can compute in linear time the treewidth of the splitgraph $S(C)$. Since there are at most n maximum cliques, this proves the theorem.   □

We now show an algorithm to compute the pathwidth of a cotriangulated graph. Let G be triangulated and let H be an interval embedding of $\overline{G}$. Hence H is triangulated and by Theorem 7.2.2 we know there exists a clique C in G such that $S(C)$ is a subgraph of H. Hence we can find the pathwidth of G by computing the pathwidth of $S(C)$ for all maximal cliques C of G and taking the minimum of those. We can make a list of all maximal cliques in $O(n^2)$ time. Using Lemma 7.2.2 we can compute the pathwidth of $S(C)$ in $O(n^\alpha)$ time, for each maximal clique C. Since there are at most n maximal cliques, the computation of the pathwidth of $\overline{G}$ can be done in $O(n^{\alpha+1})$. The algorithm can be adapted such that it produces an optimal path-decomposition within the same time bound. This proves the following theorem.

Theorem 7.2.5 *There exists a polynomial time algorithm which, given a cotriangulated graph $\overline{G}$, computes a path-decomposition of width equal to the pathwidth of $\overline{G}$.*

7.3 Convex graphs

Let $G = (X, Y, E)$ be a convex graph, with $|X| = m$, and assume the vertices of X have been ordered $1, 2, \ldots, m$ such that this ordering fulfills the adjacency property (i.e. for each $y \in Y$ the neighbors in X are consecutive). Let k be some integer. In this section we describe an $O(nk)$ algorithm which produces a path-decomposition with width at most $2k+1$ or shows that the treewidth of G is larger than k.

Consider the case where $k \geq m$. If we add edges such that X becomes a clique, then the result is a splitgraph with clique number at most $k+1$. Hence, according to lemma 7.2.1, we find a path-decomposition with width at most $k+1$. Hence we may assume without loss of generality that $k \leq m-1$.

Let Y' be the set of vertices of Y with degree at least $k+1$. Consider the subgraph G', induced by vertices of X and the vertices of Y'. We construct a path-decomposition for G' as follows. For $i = 1, \ldots, m-k$ let Z_i be the subset with:

$$Z_i \cap X = \{i, i+1, \ldots, i+k\}$$
$$Z_i \cap Y = \{y \in Y \mid Z_i \cap X \subseteq N(y)\}$$

Where $N(y)$ is the neighborhood of $y \in Y$. Notice that $Z_i \cup Y \subseteq Y'$.

Lemma 7.3.1 *Assume* $k \leq m-1$. *If the treewidth of G' is k then the pathwidth of G' is at most $2k$ and $(Z_1, \ldots, Z_{m-k})$ is a path-decomposition for G' with width at most $2k$.*

Proof. Notice that each Z_i has $k+1$ vertices of X. Each vertex $y \in Y'$ which is in Z_i is adjacent to all vertices of $Z_i \cap X$. This show there can be at most k vertices of Y in Z_i, otherwise G' has a complete bipartite subgraph $K(k+1, k+1)$, which is forbidden by lemma 2.2.1. Obviously, $(Z_1, \ldots, Z_{m-k})$ is indeed a path-decomposition. $\qquad\square$

We now show how to extend this path-decomposition to a path-decomposition for G. Consider a vertex $y \in Y$, with at most k neighbors. Notice that there exists a subset Z_i containing $N(y)$. Take such a subset Z_i containing $N(y)$ with at most $2k+1$ elements. Make a new subset $Z_{i'} = Z_i \cup \{y\}$, and make a new path-decomposition $(Z_1, \ldots, Z_i, Z_{i'}, Z_{i+1}, \ldots, Z_{m-k})$. This clearly is a path-decomposition for the subgraph of G induced by the vertices $X \cup Y' \cup \{y\}$. By induction the following lemma follows.

Lemma 7.3.2 *Let $G = (X, Y, E)$ be convex with treewidth k. Then the pathwidth of G is at most $2k+1$.*

We now show that the algorithm described above can be implemented to run in $O(nk)$ time. We assume that the vertices of X are ordered $1, \ldots, m$ such that this ordering has the adjacency property. Assume $Y = \{y_1, \ldots, y_t\}$.

Step 1 First assume that $k \geq m$. Then make a subset $Z_y = X \cup \{y\}$ for each vertex $y \in Y$. The sequence $(Z_{y_1}, Z_{y_2}, \ldots, Z_{y_t})$, (for *any* ordering $y_1, \ldots, y_t$ of the vertices of Y) is a correct path-decomposition. Stop.

Step 2 Assume $k \leq m-1$. Check if the number of edges does not exceed $nk - \frac{1}{2}k(k+1)$. If it does, then stop; the treewidth of G is larger than k.

Step 3 Otherwise, for each $y \in Y$, determine the maximum and minimum neighbor in X, say $\min(y)$ and $\max(y)$. If a vertex y has degree 0 then we set $\min(y) = 0$ and $\max(y) = -1$. The result of this step is that the set of neighbors of $y \in Y$, with degree at least one, is $\{x \in X \mid \min(y) \leq x \leq \max(y)\}$.

Step 4 Calculate for each vertex $y \in Y$ the degree, $\max(y) - \min(y) + 1$, and make a list Y' of vertices of degree at least $k + 1$.

Step 5 Initialize subsets $Z_i = \{i, \ldots, \min(m, i + k)\}$, for $i = 1, \ldots, m$. For isolated vertices of Y, we initialize $Z_0 = \emptyset$.

Step 6 For each $y \in Y'$, put y in the subsets Z_i, for $i = \min(y), \ldots, \max(y) - k$. If one of these subsets, say Z_i, gets more than $2k + 1$ vertices, then stop; the treewidth of G exceeds k. The vertices of Z_i induce a $K(k + 1, k + 1)$.

Step 7 For each $y \in Y \setminus Y'$, make a subset $Z'_y = Z_{\min(y)} \cup \{y\}$.

Step 8 For $i = 0, \ldots, m$, let $y_1^i, y_2^i, \ldots$ be the vertices $y \in Y \setminus Y'$ with $\min(y) = i$. Then the path-decomposition is $(Z_0, Z'_{y_1^0}, Z'_{y_2^0}, \ldots, Z_1, Z'_{y_1^1}, Z'_{y_2^1}, \ldots, Z_2, \ldots)$.

The discussion above proves the following theorem.

Theorem 7.3.1 *Let G be a convex graph and let k be an integer. There exists an $O(nk)$ algorithm which either determines that the treewidth exceeds k, or produces a path-decomposition of G with width at most $2k + 1$.*

7.4 Permutation graphs

In this section, let $G[\pi]$ be a permutation graph with n vertices and with treewidth k. We show there exists a path-decomposition of width at most $2k$, and we give a linear time algorithm to compute this. The algorithm outputs a set X_i for $1 \le i \le n$. A vertex j is put in all sets X_k with $\pi_j^{-1} \le k < j$ or $j \le k < \pi_j^{-1}$. The precise algorithm is given below.

```
Procedure Pathdec (input π; output X)
for i ← 1 to n do X_i ← ∅
for j ← 1 to n do
    if π_j^{-1} = j then X_j ← X_j ∪ {j}
    if π_j^{-1} > j then
        for k ← j to π_j^{-1} − 1 do X_k ← X_k ∪ {j}
    if π_j^{-1} < j then
        for k ← π_j^{-1} to j − 1 do X_k ← X_k ∪ {j}
```

For example, for the graph of figure 7.1, the computed sets are: $X_1 = \{1, 3\}$, $X_2 = \{1, 2, 3, 5\}$, $X_3 = \{2, 5\}$, $X_4 = \{4, 2, 5\}$ and $X_5 = \emptyset$. Notice that this path-decomposition is not optimal since the pathwidth of this graph is 2 and the computed path-decomposition has width 3. The next lemma shows that the constructed sets form indeed a path-decomposition.

Lemma 7.4.1 *Let $S = \{X_i \mid 1 \leq i \leq n\}$ be the subsets of vertices constructed by the algorithm. Let $P = (1, \ldots, n)$ be the path with n vertices. Then (S, P) is a path-decomposition for the permutation graph $G[\pi]$.*

Proof. We first show that each vertex is in at least one subset of S. Consider a vertex i. If $\pi_i^{-1} \geq i$ then i is in the subset X_i. If $\pi_i^{-1} < i$ then i is in the subset X_{i-1}. Notice that the subsets containing i are consecutive. The only thing left to show is that every edge is in at least one subset. Consider again a vertex i and let j be a neighbor of i. Assume without loss of generality that $i < j$. In the matching diagram, the line segment corresponding with j must intersect the line segment of i. Since $i < j$, this implies that $\pi_j^{-1} < \pi_i^{-1}$. We consider the different orderings of i, j, π_i^{-1} and π_j^{-1}. If $i < j \leq \pi_j^{-1} < \pi_i^{-1}$, then both i and j are contained in the subset X_j. If $i \leq \pi_j^{-1} \leq j \leq \pi_i^{-1}$ then both are contained in $X_{\pi_j^{-1}}$. If $i \leq \pi_j^{-1} < \pi_i^{-1} \leq j$, then both are contained in $X_{\pi_j^{-1}}$. If $\pi_j^{-1} \leq i \leq \pi_i^{-1} \leq j$, then both are contained in X_i. Finally, if $\pi_j^{-1} < \pi_i^{-1} \leq i < j$, then both i and j must be in $X_{\pi_i^{-1}}$. $\square$

We now show that the width of this path-decomposition is at most $2k$.

Lemma 7.4.2 *Each subset produced by the algorithm has at most $2k + 1$ elements.*

Proof. Consider a subset X_i. Notice that $X_i \subset S_1 \cup S_2 \cup \{i\}$ where S_1 and S_2 are defined by: $S_1 = \{j \mid j \leq i < \pi_j^{-1}\}$ and $S_2 = \{j \mid \pi_j^{-1} \leq i < j\}$. Note that, as π is a permutation, there must be as many lines in the matching diagram with their upper point left of i and their lower point right of i, as lines with their upper point right of i and their lower point left of i. Hence $|S_1| = |S_2|$. Every vertex in S_1 is adjacent to every vertex in S_2, hence the subgraph induced by $S_1 \cup S_2$ contains a complete bipartite subgraph $K(m, m)$, with $m = |S_1|$. By lemma 2.2.1, this implies that $k \geq m$. Hence $|X_i| \leq |S_1| + |S_2| + 1 \leq 2k + 1$. $\square$

Notice that the algorithm can be implemented to run in $O(nk)$ time, since at each step one new element is put into a subset. Hence we have proved the following theorem:

Theorem 7.4.1 *If $G[\pi]$ is a permutation graph with treewidth at most k, then the pathwidth of $G[\pi]$ is at most $2k$, and the $O(nk)$ time algorithm Pathdec produces a path-decomposition with width at most $2k$.*

It follows that we have $O(nk)$ approximation algorithms for pathwidth and treewidth with performance ratio 2. In chapter 9 we show that the pathwidth and treewidth are equal, and can be computed in $O(nk)$ time. Because of its simplicity and also as an introduction for the next section, we decided to include this approximation algorithm.

7.5   Cocomparability graphs

In this section, let G be a cocomparability graph with treewidth at most k. Notice that computing the exact treewidth of a cocomparability graph is NP-hard. In fact, it is shown in [4] that treewidth is NP-hard for complements of bipartite graphs.

We start with an informal discussion of the algorithm. Recall lemma 7.1.1. There exist n continous functions, $F_i : (0,1) \to R$, such that two vertices are adjacent if and only if the corresponding functions intersect. First make a vertex coloring of G, such that no two adjacent vertices have the same color. Since the treewidth of G is at most k, this can be done by using at most $k+1$ colors. Fix any position L at the line $x = 0$. This partitions the vertices of G into two sets: one set of vertices for which the corroponding functions start below L, and one set for which the corresponding functions start at position at least L. For each of these positions L we make a subset X_L as follows. For each colorclass C_t, take the top most $k+1$ functions which start below L, say $\overline{C_t}$ (take all if there are less than $k+1$). Notice that the functions corroponding with a colorclass do not intersect, hence the 'topmost functions' are well defined. For functions starting at position at least L, take those which are adjacent to $k+1$ vertices of $\overline{C_t}$, say $\mathcal{F}_t$. Notice that $\mathcal{F}_t$ is empty if $|\overline{C_t}| \leq k$. Also notice that if $|\overline{C_t}| = k+1$, then each vertex of $\mathcal{F}_t$ is adjacent to all vertices of $\overline{C_t}$. In this last case $|\mathcal{F}_t| \leq k$, otherwise there is a $K(k+1, k+1)$ subgraph. It follows that X_L has at most $(k+1)(2k+1)$ vertices, since there are at most $k+1$ color classes and for each color class C_t we have $|\overline{C_t}| + |\mathcal{F}_t| \leq 2k+1$. Notice that we only used the ordering of the functions at position $x = 0$. We can find a suitable ordering using the height function of G.

We now give the formal description of the algorithm and proof the correctness, without using the function model. We assume that a height function h of the complement $\overline{G}$ with transitive orientation F, and a coloring of G are given. Let $C_1, \ldots, C_s$ be the color classes of G, where $s = \chi(G)$ is the chromatic number of G. Since a partial k-tree can always be colored with $k+1$ colors, we may assume that the number of color classes $s \leq k+1$.

The first step of the algorithm is to renumber the vertices.

Definition 7.5.1 Let $G = (V, E)$ be a cocomparability graph with n vertices and let h be a height function of the transitively oriented complement $\overline{G}$. A height labeling of G is a bijection $L : V \to \{1, \ldots, n\}$, such that $h(x) > h(y)$ implies that $L(x) > L(y)$.

A height labeling clearly can be computed in $O(n)$ time, if the height function h is given.

Definition 7.5.2 *For* $i = 1, \ldots, n$ *and for* $t = 1, \ldots, s$, *let* $C_t(i)$ *be the set of vertices in color class* C_t *with label at most* i: $C_t(i) = C_t \cap \{x \mid L(x) \leq i\}$. *Write* $C_t(i) = \{x_1, \ldots, x_m\}$, *with* $i \geq L(x_1) > L(x_2) \ldots > L(x_m)$. *We define for* $i = 1, \ldots, n$ *and* $t = 1, \ldots, s$:

$$\overline{C_t}(i) = \begin{cases} C_t(i) & \text{if } m \leq k+1 \\ \{x_1, \ldots, x_{k+1}\} & \text{otherwise} \end{cases}$$

Notice that, since C_t is a clique in $\overline{G}$, all heights of vertices in C_t are different. It follows that $\overline{C_t}(i)$ is uniquely determined as the set of $k + 1$ vertices of $C_t(i)$ with the largest heights.

Definition 7.5.3 *For* $i = 1, \ldots, n$ *and* $t = 1, \ldots, s$, *define*

$$\mathcal{F}_t(i) = \{x \mid L(x) > i \wedge |\text{Adj}(x) \cap C_t(i)| \geq k + 1\}$$

where $\text{Adj}(x)$ *is the set of neighbors of* x.

We can now define the subsets of the path-decomposition:

Definition 7.5.4 *For* $i = 1, \ldots, n$, *let* X_i *be the following set of vertices:*

$$X_i = \bigcup_{1 \leq t \leq s} (\mathcal{F}_t(i) \cup \overline{C_t}(i))$$

In the rest of this section we prove that each subset X_i has at most $(2k + 1)\chi(G)$ elements and we show that $(X_1, \ldots, X_n)$ forms indeed a path-decomposition. The following lemma is crucial.

Lemma 7.5.1 *Each* $y \in \mathcal{F}_t(i)$ *is adjacent to all vertices of* $\overline{C_t}(i)$.

Proof. Suppose not. Let $y \in \mathcal{F}_t(i)$ be not adjacent to every vertex in $\overline{C_t}(i)$. Let $C_t(i) = \{x_1, \ldots, x_m\}$ with $h(x_1) > h(x_2) > \ldots > h(x_m)$. If $m < k + 1$, then by definition $\mathcal{F}_t(i) = \emptyset$. Hence we may assume $m \geq k + 1$ and $\overline{C_t}(i) = \{x_1, \ldots, x_{k+1}\}$. Let $x_w \in \overline{C_t}(i)$ (with $1 \leq w \leq k + 1$) be not adjacent to y. Consider the complement $\overline{G}$ with the transitive orientation F. $C_t(i)$ is a clique in $\overline{G}$ and $(x_p, x_q) \in F$ for all x_p and x_q in $C_t(i)$ with $p < q$. Since y is adjacent to x_w in $\overline{G}$, we must have $h(y) \neq h(x_w)$. Since $L(y) > L(x_w)$, it follows that $h(y) > h(x_w)$ and hence $(y, x_w) \in F$. Since F is transitive, we find that y is adjacent in $\overline{G}$ to all vertices x_p with $w \leq p \leq m$. So y can have at most $w - 1$ neighbors in $C_t(i)$ in G, hence it can not be in $\mathcal{F}_t(i)$, contradiction. $\square$

Corollary 7.5.1 *A vertex* y *with* $L(y) > i$ *is in* $\mathcal{F}_t(i)$ *if and only if it is adjacent to the vertex in* $\overline{C_t}(i)$ *with the smallest label.*

Theorem 7.5.1 *If* G *is a cocomparability graph with treewidth at most* k*, then* $|\overline{C_t}(i)| \le k+1$ *and* $|\mathcal{F}_t(i)| < k+1$.

Proof. Notice that $|\overline{C_t}(i)| \le k+1$, by definition. If $|\overline{C_t}(i)| < k+1$, then $C_t(i)$ contains less than $k+1$ elements, and then, by definition $\mathcal{F}_t(i) = \emptyset$.

Now assume that $C_t(i)$ contains at least $k+1$ vertices. Then $|\overline{C_t}(i)| = k+1$. By lemma 7.5.1 all vertices of $\mathcal{F}_t(i)$ are adjacent to all vertices of $\overline{C_t}(i)$.

The treewidth of G is at most k. Hence G can not have a complete bipartite subgraph $K(k+1, k+1)$. This implies that $|\mathcal{F}_t(i)| < k+1$.   □

Corollary 7.5.2 $\forall_i\ |X_i| \le (2k+1)\chi(G)$.

Corollary 7.5.2 shows that, if $(X_1, \ldots, X_n)$ is a path-decomposition of G, then the width of this path-decomposition is at most $2k^2 + 3k$. The next three lemmas show that $(X_1, \ldots, X_n)$ is indeed a path-decomposition.

Lemma 7.5.2 *For every vertex* x *of* G: $x \in X_{L(x)}$.

Proof. Let x be in the color class C_p. By definition, $x \in \overline{C_p}(L(x)) \subseteq X_{L(x)}$.   □

Lemma 7.5.3 *Let* $\{x, y\} \in E$ *be an edge of* G*. Then there is a subset* X_i *such that* x *and* y *are both in* X_i.

Proof. Assume $L(x) < L(y)$. Notice that x and y are not in the same color class, since they are adjacent. Consider the color class of x, say $C_p = \{x_1, \ldots, x_m\}$, and let $L(x_1) > L(x_2) > \ldots > L(x_m)$. Let $x = x_j$ for some $1 \le j \le m$. Clearly, if $j \le k+1$, then x and y are both contained in $X_{L(y)}$ since $x \in \overline{C_p}(L(y)) \subseteq X_{L(y)}$. Now assume $j > k+1$. Consider $z = x_{j-k}$. If $L(y) < L(z)$ then x and y are both contained in $X_{L(y)}$, since $x \in \overline{C_p}(L(y))$. If $L(y) > L(z)$, then $x \in \overline{C_p}(L(z))$ and $y \in \mathcal{F}_p(L(z))$.   □

Lemma 7.5.4 *The subsets* X_i *containing a given vertex* x*, are a consecutive subsequence of* $(X_1, \ldots, X_n)$.

Proof. Assume $L_1 < L_2$ and $x \in X_{L_1} \cap X_{L_2}$. Let $L_1 < L < L_2$. We prove that $x \in X_L$. We consider three cases:

1. $L(x) \ge L_2$. Since $x \in X_{L_1}$, x must be adjacent to at least $k+1$ vertices of some color class which are in X_{L_1}. Clearly, this also holds for every $L_1 \le L < L(x)$. Hence $x \in X_L$.

2. $L(x) \leq L_1$. Since $x \in X_{L_2}$, x must be among the vertices in its color class with the $k+1$ largest heights which are in X_{L_2}. But, then clearly this must hold for every $L(x) \leq L \leq L_2$.

3. $L_1 < L(x) < L_2$. The argument of the first case shows that $x \in X_L$ for all $L_1 \leq L < L(x)$. In the second case it is shown that $x \in X_L$ for all $L(x) \leq L \leq L_2$.

$\square$

By lemmas 7.5.2, 7.5.3, and 7.5.4 the sequence of subsets $(X_1, \ldots, X_n)$ is a path-decomposition for the cocomparability graph G. According to corollary 7.5.2, the width of this path-decomposition is at most $(2k+1)\chi(G) - 1$. In the following theorem we summarize these results.

Theorem 7.5.2 *Let G be a cocomparability graph. The sequence $(X_1, \ldots, X_n)$ with X_i defined in definition 7.5.4, is a path-decomposition for G. If the treewidth of G is at most k, then the width of this path-decomposition is at most $(2k+1)\chi(G) - 1 \leq 2k^2 + 3k$.*

Consider the time it takes to compute the sets X_i. We can sort all the color classes according to increasing labels in $O(nk)$ time. Then each set $\overline{C}_t(i)$ can be computed in $O(k)$ time. Now notice that $\mathcal{F}_t(i) \subseteq \mathcal{F}_t(i+1) \cup \{x \mid L(x) = i+1\}$. If we use an adjacency matrix to represent G, we can compute each set $\mathcal{F}_t(i)$ in time $O(k)$: An element y in $\mathcal{F}_t(i+1) \cup \{x \mid L(x) = i+1\}$ is in $\mathcal{F}_t(i)$ if and only if y is adjacent to the vertex with the smallest label in $\overline{C}_t(i)$ (corollary 7.5.1). The sets $\overline{C}_t(i)$ and $\mathcal{F}_t(i)$ have at most $k+1$ elements and there are at most $k+1$ of each for each i. Hence we can easily compute each set X_i in time $O(k^2)$.

Corollary 7.5.3 *Let G be a cocomparability graph with treewidth k. Assume a vertex coloring of G with at most $k+1$ colors, and a transitive orientation F of the complement $\overline{G}$ are given. If an adjacency matrix is used to represent G, then a path-decomposition of G with width at most $2k^2 + 3k$ can be computed in time $O(nk^2)$.*

Chapter 8

Treewidth of chordal bipartite graphs

The chordal bipartite graphs form a large class of perfect graphs containing, for example, not only the convex and biconvex bipartite graphs, but also the bipartite permutation graphs and the bipartite distance-hereditary graphs (or $(6,2)$-chordal bipartite graphs). Many NP-hard problems remain NP-hard when restricted to the class of chordal bipartite graphs. For example HAMILTONIAN CYCLE, HAMILTONIAN PATH, DOMINATING SET, CONNECTED DOMINATING SET, INDEPENDENT DOMINATING SET, and STEINER TREE [130, 53]. The recognition problem for chordal bipartite graphs can be solved in $O(\min(m \log n, n^2))$ [122, 156]. Since so many NP-hard problems remain NP-hard when restricted to chordal bipartite graphs, it is of importance to be able to use the partial k-tree algorithms for these problems. In this chapter we give a polynomial time algorithm to find the treewidth and a tree-decomposition of optimal width for a chordal bipartite graph.

We do not claim that our algorithm for the treewidth of chordal bipartite graphs is a very practical one, but we feel that it is one of the first non-trivial polynomial time algorithms for computing the treewidth of a relatively large class of graphs. Note that it considerably narrows the gap between classes where treewidth is computable in polynomial time and the classes for which the corresponding decision problem is NP-complete.

8.1 Preliminaries

We start with some definitions and easy lemmas. For more information the reader is referred to [76] or [38].

Definition 8.1.1 *A graph is called* chordal bipartite *(or weakly chordal bipartite) if it is bipartite and each cycle of length at least six has a chord.*

A chord (x, y) in a cycle C of even length is *odd* if the distance between x and y in the cycle is odd.

Definition 8.1.2 *A graph is called* strongly chordal *if it is chordal and each cycle of even length at least six has an odd chord.*

Definition 8.1.3 *For a bipartite graph* $G = (X, Y, E)$ *let* $\mathrm{split}(G) = (X, Y, \hat{E})$ *with* $\hat{E} = E \cup \{(x, x') \mid x, x' \in X \wedge x \neq x'\}$.

A split graph is a graph for which there is a partition of the vertices in two sets such that one set induces an independent set and the other set induces a clique. These are exactly the graph which are both triangulated and cotriangulated. The following characterization of chordal bipartite graphs appeared in [52].

Lemma 8.1.1 $G = (X, Y, E)$ *is chordal bipartite if and only if* $\mathrm{split}(G)$ *is strongly chordal.*

If x is a vertex of a graph $G = (V, E)$, we denote by $N[x]$ the *closed neighborhood* of x, i.e. $N[x] = N(x) \cup \{x\} = \{y \mid y = x \text{ or } (x, y) \in E\}$.

Definition 8.1.4 *A vertex* v *is* simple *if for all* $x, y \in N[v]$, $N[x] \subseteq N[y]$ *or* $N[y] \subseteq N[x]$.

Notice that a simple vertex is simplicial (i.e., the neighborhood is complete). We shall use the following property of strongly chordal graphs [60].

Lemma 8.1.2 *A graph* G *is strongly chordal if and only if every induced subgraph has a simple vertex.*

Definition 8.1.5 *Let* $G = (X, Y, E)$ *be a bipartite graph. Then* $(u, v) \in E$ *is called a* bisimplicial edge *if* $N(u) \cup N(v)$ *induces a complete bipartite subgraph of* G.

Definition 8.1.6 *Let* $G = (X, Y, E)$ *be a bipartite graph. Let* $(e_1, \ldots, e_k)$ *be an ordering of the edges of* G. *For* $i = 0, \ldots, k$ *define the subgraph* $G_i = (X_i, Y_i, E_i)$ *as follows:* $G_0 = G$, *and for* $i \geq 1$ G_i *is the subgraph of* G_{i-1} *with* $X_i = X_{i-1}$, $Y_i = Y_{i-1}$ *and* $E_i = E_{i-1} \setminus \{e_i\}$ *(i.e., the edge* e_i *is removed but not the end vertices). The ordering* $(e_1, \ldots, e_k)$ *is a* perfect edge-without-vertex elimination ordering *for* G *if each edge* e_i *is bisimplicial in* G_i, *and* G_k *has no edge.*

The following lemma appears for example in [38].

Lemma 8.1.3 G *is chordal bipartite if and only if there is a perfect edge-without-vertex elimination ordering.*

The following lemma implies that we can start a perfect edge-without-vertex elimination ordering with *any* bisimplicial edge.

Lemma 8.1.4 *Let G be chordal bipartite. Let e be a bisimplicial edge in G. Let G′ be the graph obtained from G by deleting the edge e but not the end vertices of e. Then G′ is chordal bipartite.*

Proof. Assume G′ has a chordless cycle C of length ≥ 6. Let $e = (x, y)$. Then, clearly, x and y must be elements of C. The neighbors of x and y in the cycle form a square (i.e., a chordless 4-cycle). This shows that C cannot be chordless in G′. □

In [74] it is shown that a bisimplicial edge in a chordal bipartite graph with n vertices can be found in $O(n^2)$ time.

Corollary 8.1.1 *A perfect edge-without-vertex elimination scheme in a chordal bipartite graph can be determined in time $O(n^2 m)$, where n is the number of vertices and m is the number of edges of the graph.*

8.2 Triangulating chordal bipartite graphs

In this section, let $G = (X, Y, E)$ be chordal bipartite. We give a method to triangulate G and prove the correctness. We denote complete bipartite subgraphs as $M = (A, B)$, i.e., the vertex set of this graph M is $A \cup B$ and the edge set $E = \{(a, b) \mid a \in A \wedge b \in B\}$. In this chapter we require by definition that a complete bipartite graph (A, B) is such that $|A| \geq 2$ and $|B| \geq 2$. If $G = (X, Y, E)$ is a bipartite graph, then we call the sets X and Y the color classes of G.

Lemma 8.2.1 *If $G = (X, Y, E)$ is chordal bipartite, then it contains at most $|E|$ maximal complete bipartite subgraphs.*

Proof. G is chordal bipartite, hence there is a perfect edge-without-vertex elimination ordering $(e_1, \ldots, e_k)$. Consider a maximal complete bipartite subgraph, (A, B). Let e_i be the first edge in the ordering which is an edge of (A, B). Let $e_i = (x, y)$ with $x \in A$ and $y \in B$. Since e_i is bisimplicial and (A, B) is maximal we have $A = N(y)$ and $B = N(x)$. Thus the maximal complete bipartite subgraph (A, B) is completely and uniquely determined by e_i. This proves the lemma. □

Remark. It is not difficult to see that there exist chordal bipartite graphs for which the number of maximal complete bipartite subgraphs is $\Omega(n^2)$.

If (A, B) is a complete bipartite graph and H is a triangulation, then either H[A] or H[B] is a complete subgraph of H (otherwise there would be a chordless square; see Lemma 2.2.1, page 18). Now let G be chordal bipartite, and let $\mathcal{M}$ be the set of maximal complete bipartite subgraphs (A, B) of G with $|A| \geq 2$ and $|B| \geq 2$. If H is a triangulation of G, then for each $(A, B) \in \mathcal{M}$, either $H[A]$ or $H[B]$ is a complete subgraph of H. Consider the following process. For *each* $(A, B) \in \mathcal{M}$, choose one color class $C \in \{A, B\}$, and add all edges between vertices of C. We say the color class C is completed. The following example shows, that the resulting graph need not be chordal.

Example. Take the K(4,4) with color classes $A = \{a, b, c, d\}$ and $B = \{p, q, r, s\}$ and delete the edges (p, d) and (a, s). Call this graph G. It is easy to see that G is chordal bipartite (it is even bipartite permutation). For the maximal complete bipartite subgraph $(\{a, b, c, d\}, \{q, r\})$ we choose the color class $\{a, b, c, d\}$ and change this into a clique. For the maximal complete bipartite subgraph $(\{b, c\}, \{p, q, r, s\})$ we choose the color class $\{p, q, r, s\}$ and make this complete. The resulting graph is not triangulated, because there is a chordless square induced by $\{a, d, p, s\}$.

Definition 8.2.1 *Let* $M_1 = (A_1, B_1)$ *and* $M_2 = (A_2, B_2)$ *be two maximal complete bipartite subgraphs of a graph G. We say that* M_1 *and* M_2 *cross if either* $A_2 \subset A_1$ *and* $B_1 \subset B_2$, *or* $A_1 \subset A_2$ *and* $B_2 \subset B_1$.

In the example above, the maximal complete bipartite subgraphs $(\{b, c\}, \{p, q, r, s\})$ and $(\{a, b, c, d\}, \{q, r\})$ cross.

Definition 8.2.2 *For each* $M \in \mathcal{M}$ *let* $C(M)$ *be one of the color classes. The set* $\mathcal{C} = \{C(M) \mid M \in \mathcal{M}\}$ *is called* feasible, *if for each pair* $(A_1, B_1), (A_2, B_2) \in \mathcal{M}$ *that cross with* $A_2 \subseteq A_1$ *and* $B_1 \subseteq B_2$, *not both* A_1 *and* B_2 *are in* $\mathcal{C}$.

We want to prove in this section that if $\mathcal{C}$ is feasible and we complete each $C \in \mathcal{C}$, then the resulting graph is triangulated. Notice that there exists a feasible set of color classes: simply complete for each $M \in \mathcal{M}$ the smallest color class. We give an example showing that this needs not be optimal. The (convex) graph of Figure 8.1 (page 97) has two maximal complete bipartite subgraphs, $(\{p, q, r, s\}, \{c, d, e\})$ and $(\{p, q, r\}, \{a, b, c, d, e\})$. If we complete for both the smaller color class we obtain a triangulated graph with a maximal clique $\{p, q, r, c, d, e\}$ of six vertices. Another possibility is that we complete $\{p, q, r, s\}$. In this case the largest clique has only five vertices.

Theorem 8.2.1 *Let* $\mathcal{C}$ *be a feasible set of color classes of a chordal bipartite graph G. Let H be the graph obtained by making each* $C \in \mathcal{C}$ *complete. Then H is triangulated.*

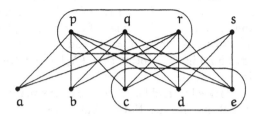

Figure 8.1: feasible set need not be optimal

Proof. Assume by way of contradiction that G is a *minimal* counterexample. Thus for every induced subgraph the theorem is true. By assumption H is not chordal. Let S be a chordless cycle of length greater than three in H. Let $G = (X, Y, E)$. We call the vertices of X red and the vertices of Y black. An edge of S is called red (black), if *both* its end vertices are red (black). The red and black edges of S are called colored edges. Notice that there must exist at least one colored edge, otherwise the cycle would also be a chordless cycle in G, and since G is chordal bipartite, S must be a square. But a square is contained in some maximal complete bipartite subgraph. Hence at least one of the color classes must have been completed. Let $\mathcal{R}$ be the set of all red edges of S and $\mathcal{B}$ be the set of all black edges of S. Without loss of generality we may assume that $\mathcal{R} \neq \emptyset$.

Consider a red vertex $x \notin S$, and let $e \in \mathcal{B}$ be a black edge of S. We say that x *creates* e, if there is a maximal complete bipartite subgraph (A, B) with $e \subseteq B$, $x \in A$, and $B \in C$. A red vertex r is called *redundant*, if $r \notin S$ and it does not create any black edge. A black vertex $b \notin S$ is called redundant, if it does not create a red edge.

Claim *There are no redundant vertices.*

Proof of claim. Suppose there is a redundant red vertex x. Consider the induced subgraph $G' = G[V \setminus \{x\}]$. Clearly, also G' is chordal bipartite. We create a feasible set of color classes C' as follows. Consider a maximal complete bipartite subgraph $M = (A, B)$ of G'. If M is also maximal in G, we put $C(M)$ in C'. If M is not maximal in G then $N = (A \cup \{x\}, B)$ must be a maximal complete bipartite subgraph of G. In this case we put $C(N) \setminus \{x\}$ in C'. It is easy to see that C' is a feasible set of color classes for G'.

Let H' be the graph obtained from G' by completing each set of C'. We claim that S is also a chordless cycle in H'.

Let e be a colored red edge of S. There is a maximal complete bipartite subgraph (A, B), with $e \subseteq A$ and $A \in C$. If $x \notin A$, then (A, B) is a maximal complete bipartite subgraph of G'. Hence $A \in C'$. Now assume $x \in A$. Then $|A| \geq 3$. Take B' such that $B \subseteq B'$ and $(A \setminus \{x\}, B')$ is a maximal complete bipartite subgraph of G' (i.e. B' is the set of common neighbors of $A \setminus \{x\}$). If (A, B') is maximal in G, then $B' = B$ and $A \setminus \{x\} \in C'$. Otherwise $(A \setminus \{x\}, B')$ is maximal in G. Notice that (A, B) and $(A \setminus \{x\}, B')$ cross. Hence $A \setminus \{x\}$ is in C and hence also in C'.

Now let e be a black edge, and let (A, B) be a maximal complete bipartite subgraph with $e \subseteq B$ and $B \in C$. Notice that since x is redundant, $x \notin A$. Hence (A, B) is also maximal in G' and $B \in C'$. This shows that S is also a cycle in G'.

Since G was a minimal counterexample, there must be a chord in G' between two vertices p and q of S. Without loss of generality assume p and q are both red (the case where they are both black is similar). There exists a maximal complete bipartite subgraph (A, B) in G' such that $p, q \in A$ and $A \in C'$. By definition either $A \cup \{x\} \in C$ or $A \in C$. Hence (p, q) is also an edge in G. This proves that there are no redundant red vertices. A similar argument show that there are no redundant black vertices either. $\square$

Consider the graph $G^* = \text{split}(G)$ obtained by making Y complete. Since G is chordal bipartite, we know that G^* is strongly chordal. Hence we know that G^* has a simple vertex. Let s be the simple vertex. The next lemma shows that s must be a red vertex.

Claim *A simple vertex s of G^* is red.*

Proof of claim. Assume $s \in Y$ is simple. If s is not an element of the cycle, it must be in some maximal complete bipartite subgraph (A, B) creating some edge, otherwise s would be redundant. But then it has at least two nonadjacent neighbors, which is a contradiction. Assume s is a black vertex of the cycle S. Clearly s can not have two red neighbors, because red vertices are not adjacent in G. Hence s is incident with at least one edge of $\mathcal{B}$. Consider a maximal path of S, which contains only black vertices, and which contains s. Let p and q, be the end vertices of this path. We know that $p \neq q$. Since s is simple, and p and q are neighbors of s in G^*, either $N[p] \subseteq N[q]$ or $N[q] \subseteq N[p]$. Since p and q are both incident with a red vertex in S, it follows that S contains only one red vertex. This is a contradiction, because we assumed that $\mathcal{R} \neq \emptyset$. $\square$

Claim *A simple vertex s is an element of S.*

Proof of claim. We know s is red. Assume s is not in S. Then we know that s creates some black edge $e = (p, q)$. Without loss of generality assume $N[p] \subseteq N[q]$. If p has a red neighbor in S, then this would also be a neighbor

of q, and there would be a chord in S. Hence p has another black neighbor r
in S. Consider the maximal complete bipartite subgraph (K, L) with $p, r \in L$
and $L \in C$. Then p is adjacent to every vertex of K. Hence also q is adjacent
to every vertex of K. It follows that $q \in L$. But $L \in C$. Hence q and r are
adjacent in H. This is a chord in S.   □

We have to consider three more cases. First we show that s is incident with
at least one edge of $\mathcal{R}$.

Claim *A simple vertex s is incident with at least one red edge of S.*

Proof of claim. We know s is a red vertex of S. Assume it has two black
neighbors in S, x and y. We may assume $N[x] \subseteq N[y]$. If in the cycle x
has two red neighbors, then also y is adjacent to both red neighbors, which
is a contradiction. Hence x is adjacent to a black vertex z in S. Consider a
maximal complete bipartite subgraph (K, L) creating the edge (x, z). Then
x is adjacent to every vertex of K, and hence this also holds for y. But then
$y \in L$. $L \in C$ hence x and y are adjacent in H. This is a chord in S.   □

Claim *A simple vertex s is incident with exactly one red edge of S.*

Proof of claim. We know that $s \in S$ and s has at least one red neighbor in
S. Assume s has *two* red neighbors x and y in S. Consider the maximal
complete bipartite subgraphs (K_x, L_x) creating the edge (s, x) and (K_y, L_y)
creating (s, y). Now there must exist $p \in L_x \setminus L_y$ which is not adjacent to
y, otherwise $(K_x \cup \{y\}, L_x)$ is complete bipartite. Also there must exist $q \in
L_y \setminus L_x$ which is not adjacent to x. Without loss of generality we may assume
$N[p] \subseteq N[q]$. But then we have a contradiction since $x \in N[p] \subseteq N[q]$.   □

Hence we know that a simple vertex s is a red vertex of the cycle S with
one red neighbor x and one black neighbor y. Let (K_x, L_x) be the maximal
complete bipartite subgraph that creates (s, x). Since s is simple and y is
not adjacent to x, we must have $N[y] \cap X \subseteq K_x$. Now assume that in the
cycle y is adjacent to another red vertex u. Then $u \in N[y] \cap X \subseteq K_x$. But
then u, x and s would form a triangle, which is a contradiction. Hence y
is adjacent to a black vertex v in the cycle. Let (K_v, L_v) be the maximal
complete bipartite subgraph creating (y, v). Notice that $K_v \subseteq N[y] \cap X \subseteq K_x$.
It follows that $L_x \subseteq L_v$, and hence (K_x, L_x) and (K_v, L_v) cross. But L_v and
K_x are completed, hence C is not feasible. This completes the proof of
Theorem 8.2.1.   □

In the next section we show that there exists a feasible set of color classes
of G such that the triangulated graph H is optimal, i.e. has the smallest
possible clique number.

8.3 Optimal triangulations of chordal bipartite graphs

Let $G = (X, Y, E)$ be a chordal bipartite graph. If $\mathcal{C}$ is a feasible set of color classes, we denote by $H_{\mathcal{C}}$ the chordal graph obtained from G by completing each $C \in \mathcal{C}$. In this section we show the following. There is a feasible set of color classes $\mathcal{C}$, such that the triangulation $H_{\mathcal{C}}$ minimizes the clique number.

Theorem 8.3.1 *Let G be a chordal bipartite graph with treewidth $\leq k$. Then there exists a feasible set of color classes $\mathcal{C}$ such that $H_{\mathcal{C}}$ has clique number $\leq k + 1$.*

Proof. Let H be any triangulation of G. We show that there is a feasible set of color classes $\mathcal{C}$ such that $H_{\mathcal{C}}$ is a subgraph of H. Let $\mathcal{M}$ be the set of all maximal complete bipartite subgraphs of G. Suppose we list the elements of $\mathcal{M}$ one by one, creating a feasible set of color classes $\mathcal{C}$ as follows. Start with $\mathcal{C} = \emptyset$. For each $M \in \mathcal{M}$ at least one of the color classes is a complete subgraph in H. Let $M = (A, B)$ and assume A is complete in H. Check the list $\mathcal{C}$ created thus far, if there is a maximal complete subgraph (C, D) which crosses with (A, B), such that $C \subseteq A$ and $D \in \mathcal{C}$. If not, then put A in $\mathcal{C}$. Otherwise put B in $\mathcal{C}$. Notice that there cannot be a maximal complete bipartite subgraph (K, L) crossing with (A, B), such that $L \subseteq B$ and $K \in \mathcal{C}$. Otherwise (K, L) and (C, D) would also cross with $C \subseteq K$ and $K, D \in \mathcal{C}$. This shows that in this way we create a feasible set of color classes. Now notice that $H_{\mathcal{C}}$ is a subgraph of H, hence the clique number of $H_{\mathcal{C}}$ does not exceed the clique number of H.   □

8.4 Cliques in optimal triangulations of chordal bipartite graphs

Let $G = (X, Y, E)$ be a chordal bipartite graph and let $\mathcal{C}$ be a feasible set of color classes. In this section we analyze the structure of the maximal cliques in $H_{\mathcal{C}}$. We first look at a clique B with vertices only in one color class.

Definition 8.4.1 *Let B be a clique with all vertices in Y. We say that a set $\mathcal{A}$ of maximal complete bipartite subgraphs is a cover for B if the following two conditions are satisfied:*

1. *for each $(A', B') \in \mathcal{A}$: $B' \in \mathcal{C}$, and*

2. *for every pair $x, y \in B$ there is a maximal complete bipartite subgraph $(A', B') \in \mathcal{A}$ such that $x, y \in B'$.*

Theorem 8.4.1 *Let* B *be a clique of* H_C *with all vertices in* Y. *Assume* $|B| \geq 2$. *A minimal cover for* B *has only one element.*

Proof. Let $\mathcal{A} = \{(A_1, B_1), \ldots (A_t, B_t)\}$ be a minimal cover for B. Assume $t \geq 2$. Let $A = \bigcup_i A_i$.

Claim *Every vertex of* A *has at least two neighbors in* B.

Proof of claim. Assume there is a vertex $a \in A$ with only one neighbor b in B. There is a complete bipartite subgraph (A_i, B_i) containing a. We claim that B is covered by $\mathcal{A} \setminus \{(A_i, B_i)\}$. For every other vertex $b' \in B$ the edge (b, b') must be in some (A_j, B_j) with $j \neq i$. This proves the lemma. $\square$

Claim *Every vertex of* B *has at least two neighbors in* A.

Proof of claim. Let $b \in B$. Take another vertex $b' \in B$. The edge (b, b') must be in some maximal complete bipartite subgraph (A_i, B_i). $\square$

Take the subgraph H of G induced by $A \cup B$. Let $W = \text{split}(H)$ obtained by making A a complete graph. W is strongly chordal. Each vertex of A has two nonadjacent neighbors, hence A cannot contain a simple vertex. Let $b \in B$ be a simple vertex. Let $a_1, \ldots, a_k$ be the neighbors of b in A, and let $N[a_1] \subseteq N[a_2] \subseteq \ldots \subseteq N[a_k]$. Notice that $k \geq 2$. Since A is a clique in W, we have that $N[a_1] \cap B \subseteq \ldots \subseteq N[a_k] \cap B$. We claim that not all a_i's can be contained in one color class A_i.

Claim *None of the color classes* A_i *contains all neighbors* $a_1, \ldots, a_k$ *of* b *in* A.

Proof of claim. Assume that $\{a_1, \ldots, a_k\} \subseteq A_i$ for some color class A_i. Take $b' \notin B_i$. There exists a maximal complete bipartite subgraph (A_j, B_j) such that $\{b, b'\} \subseteq B_j$. Clearly $A_j \subseteq \{a_1, \ldots, a_k\} \subseteq A_i$. This implies $B_i \subseteq B_j$. This is a contradiction since $\mathcal{A}$ is minimal. $\square$

Now we can finish the proof of Theorem 8.4.1. Let (A_1, B_1) be a maximal complete bipartite subgraph containing a_1. Let $i > 1$. Notice $B_1 \subseteq N[a_1] \cap B \subseteq N[a_i] \cap B$. Hence $(A_1 \cup \{a_i\}, B_1)$ is a complete bipartite subgraph. This implies that $a_i \in A_1$ for all $1 \leq i \leq k$, which is a contradiction. This proves Theorem 8.4.1. $\square$

The following theorem shows that the structure of the maximal cliques in H_C is very simple.

Theorem 8.4.2 *Let* K *be a maximal clique in* H_C *with* $|K| > 2$. *Let* $K_x = K \cap X$ *and* $K_y = K \cap Y$. *Assume* $|K_x| \geq 2$. *Then one of the following two cases holds:*

1. $|K_y| = 1$ *and there exists a maximal complete bipartite subgraph* (A, B) *such that* $K_x = A$, $y \in B$ *and* $A \in C$, *or*

2. $|K_y| > 1$ *and there exist maximal complete bipartite subgraphs* (A_1, B_1) *and* (A_2, B_2), *with* $A_1 \in C$ *and* $B_2 \in C$ *such that* $K_x \subseteq A_1$ *and* $K_y \subseteq B_2$.

Proof. By Theorem 8.4.1 there exists a maximal complete bipartite subgraph (A_1, B_1) such that $K_x \subseteq A_1$ and $A_1 \in C$. Assume $K_y = \emptyset$. Then clearly, K cannot be maximal, since for any vertex $y \in B_1$, $A_1 \cup \{y\}$ is a clique. Assume $|K_y| = 1$. Let $K_y = \{y\}$. Notice that $(K_x, \{y\})$ is contained in a complete bipartite subgraph $(K_x, B_1 \cup \{y\})$ in G. Hence $(K_x, B_1 \cup \{y\})$ is contained in a maximal complete bipartite subgraph (A_2, B_2) in G. Since $B_1 \subseteq B_2$, we have $A_2 \subseteq A_1$. Since C is feasible, and $A_1 \in C$, we must have $A_2 \in C$. Hence the first case holds. Now assume $|K_y| \geq 2$. By Theorem 8.4.1 there exists a maximal complete bipartite subgraph (A_2, B_2) with $K_y \subseteq B_2$ and $B_2 \in C$.

This proves the theorem.   □

8.5   An algorithm for the treewidth of chordal bipartite graphs

Let $G = (X, Y, E)$ be a chordal bipartite graph. In this section we show a polynomial time algorithm which outputs either a valid statement that the treewidth of G exceeds k, or a triangulation of G such that the clique number does not exceed $k + 1$. We first describe the algorithm. In the method we use, we construct a digraph with vertex set $\mathcal{M}$. We direct an edge from $M_1 = (A_1, B_1)$ to $M_2 = (A_2, B_2)$, with $A_1, A_2 \in X$, if $A_1 \in C$ then necessarily also $A_2 \in C$. Next we color some of the vertices of this graph red or black. A vertex (A, B) is colored red if necessarily $A \in C$ and black if necessarily $B \in C$. In the next step we try to extend this coloring to the whole graph.

step 1 If $k \leq 3$ then use the algorithm described in [124] to either decide that the treewidth of G exceeds k or, to find a suitable triangulation. If $k > 3$ perform the following steps.

step 2 First make a list $\mathcal{M}$ of all maximal complete bipartite subgraphs of G.

step 3 Make a directed graph W with vertex set $\mathcal{M}$ as follows. Let $M_1 = (A_1, B_1)$ and $M_2 = (A_2, B_2)$ be two different elements of $\mathcal{M}$. We direct an edge from M_1 to M_2 if one of the following cases holds:

- Compute the maximal number of vertices of a maximal complete bipartite subgraph in the induced subgraph $G[A_1 \cup B_2]$. (M_1, M_2) is a directed edge if this size exceeds $k + 1$.

- If M_1 and M_2 cross, with $A_2 \subseteq A_1$ then (M_1, M_2) is a directed edge.

step 4 Color some of the vertices of W as follows. If $M = (A, B)$ is a maximal complete bipartite subgraph with $|A| > k$ and $|B| > k$ then output that the treewidth of G exceeds k. Otherwise, if $|B| > k$, then we color M red and if $|A| > k$, then we color M black. In case $|A| \leq k$ and $|B| \leq k$, then we do not color M yet.

step 5 While there is some arc (M_1, M_2) with M_1 colored red and M_2 not colored red then consider the following cases:

- If M_2 is black then output that the treewidth of G exceeds k

- If M_2 is not colored, then color M_2 red.

step 6 All vertices which do not yet have a color, are colored black.

step 7 For all elements $M = (A, B)$ of $\mathcal{M}$: If the color of M is red, then complete A, and if the color of M is black then complete B.

Theorem 8.5.1 *Let G be chordal bipartite and let k be an integer. The treewidth of G $\leq k$ if and only if the algorithm produces a triangulation with clique number at most $k + 1$.*

Proof. Clearly, the theorem holds when $k \leq 3$. Assume $k > 3$. Clearly, if the algorithm does not produces a triangulation, then some necessary condition is not satisfied and hence the treewidth is more than k. Assume each $M \in \mathcal{M}$ is colored. Consider the set $\mathcal{C}$ of color classes, defined as follows. If $M = (A, B)$ is red, then $A \in \mathcal{C}$, otherwise $B \in \mathcal{C}$. It is easily checked that $\mathcal{C}$ is a feasible set of color classes. It follows from Theorem 8.2.1 that the algorithm produces a triangulation H. Assume H has a maximal clique K with more than $k + 1$ vertices. By Theorem 8.4.2 there are two cases to consider. First consider the case $|K_y| = 1$, let $K_y = \{y\}$. Then $|K_x| > k$. There is a maximal complete bipartite subgraph $M = (A, B)$, with $K_x \subseteq A \in \mathcal{C}$, and $y \in B$. However, the algorithm can not color M red (step 4) hence A cannot be in $\mathcal{C}$. A similar argument shows that $|K_x| = 1$ is also not possible. Now assume $|K_x| \geq 2$ and $|K_y| \geq 2$. By Theorem 8.4.2, there exist complete maximal bipartite subgraphs, $M_1 = (A_1, B_1)$ and $M_2 = (A_2, B_2)$, such that $K_x \subseteq A_1$

and $K_y \subseteq B_2$, with $A_1 \in \mathcal{C}$ and $B_2 \in \mathcal{C}$. This means that M_1 is colored red and M_2 is colored black. But (K_x, K_y) is a complete bipartite subgraph in the induced subgraph $G[A_1 \cup B_2]$, hence there is an arc (M_1, M_2) (step 3). This is a contradiction.   □

In the last part of this section we discuss the running time of the algorithm and we show it is polynomial. We assume the graph is connected. Consider the time it takes to find all maximal complete bipartite subgraphs. Corollary 8.1.1 shows we can find a perfect edge without vertex elimination ordering in time $O(n^2 m)$. Each bisimplicial edge gives a complete bipartite subgraph. It follows that we can find a list with at most m complete bipartite subgraph in time $O(m^2)$.

Given a list of at most m complete bipartite subgraphs. How fast can we find the maximal ones? We concentrate on one color classes for each complete bipartite subgraph first. We have to indicate for each pair of color classes whether or not one is contained in the other. Let subsets A_i, $i = 1, \ldots, m$ of $\{1, \ldots, n\}$ be given. We want to find out for each pair whether one is contained in the other or not. Consider the $n \times m$ 0/1-matrix incidence B, with rows corresponding to $1, \ldots, n$ and collums corresponding to $A_1, \ldots, A_m$. Compute $B^T B$ in time $O(m^{2.38})$ ([158]), (assuming the graph is connected, i.e., $m > n$). Notice that one of A_i and A_j is contained in the other one iff $(B^T B)_{i,j} = \min(|A_i|, |A_j|)$. (since $(B^T B)_{i,j} = |A_i \cap A_j|$.)

Hence we can decide this indeed in time $O(m^{2.38})$ for al pairs. From this list (of m complete bip subgraphs) we can get the maximal complete bipartite subgraphs in time $O(m^{2.38})$.

Now clearly, the time complexity is dominated by step 3 of the algorithm. This step can be performed efficiently as follows. Let $M_i = (A_i, B_i)$ ($1 \leq i \leq t$) be an arbitrary ordering of the maximal complete bipartite subgraph ($t = |\mathcal{M}|$). We show how to compute for each pair M_i and M_j the maximum size of a complete bipartite subgraph in $G[A_i \cup B_j]$. Notice that this is equal to $\max_{1 \leq k \leq t}(|A_i \cap A_k| + |B_k \cap B_j|)$. First construct matrices P and Q with $P_{xy} = |A_x \cap A_y|$ and $Q_{xy} = |B_x \cap B_y|$. Then compute the product $R = P \otimes Q$, which is defined by $R_{xy} = \max_{1 \leq k \leq t}(P_{xk} + Q_{ky})$. In [146] it is shown that this product can also be computed in $O(t^\alpha)$ which is the time needed to compute the ordinary product of two $t \times t$ matrices (i.e., $\alpha < 2.38$ [158]). It follows that we can construct the digraph W of step 3 within this time. This proves the following theorem.

Theorem 8.5.2 *Let G be chordal bipartite and let k be an integer. Then there is a polynomial time algorithm to decide whether the treewidth of G is at most k. If so, the algorithm returns a chordal embedding of G with cliquesize at most $k + 1$. The algorithm can be implemented to run in time $O(m^\alpha)$ (assuming the graph is connected) which is the time needed to multiply two $m \times m$ matrices.*

Chapter 9

Treewidth and pathwidth of permutation graphs

We showed in section 7.4 that the pathwidth of a permutation graph is at most two times the treewidth of that graph, and we gave a linear time algorithm which produces a path-decomposition of width within this bound. In this chapter, we show that for permutation graphs, the treewidth and pathwidth are in fact equal, and we give an algorithm which computes the (exact) treewidth. Our algorithm to decide whether the treewidth (pathwidth) is at most some given integer k, can be implemented to run in $O(nk)$ time (when the matching diagram is given). We show that this algorithm can easily be adapted to compute the treewidth of a permutation graph in $O(nk)$ time, where k is the actual treewidth.

Permutation graphs have many applications in scheduling problems. See for example [59] where permutation graphs are used to describe the memory requirements of a number of programs at a certain time (see also [76]). Permutation graphs also arise in a very natural way in the problem of sorting a permutation, using queues in parallel. In [76] it is shown that this problem is closely related to the coloring problem of permutation graphs. Other applications occur for example in VLSI layout (see e.g. [152]). There is a long list of papers, which mainly appeared in the last ten years, studying the algorithmic complexity of NP-hard graph problems when restricted to permutation graphs. Indeed, there exist fast algorithms for many NP-hard problems like MAX WEIGHT CLIQUE, MAX WEIGHT INDEPENDENT SET, FEED-BACK VERTEX SET and DOMINATING SET when restricted to permutation graphs [39, 40, 43, 61, 76]. However some problems remain NP-hard, like COCHROMATIC NUMBER [73, 170], and ACHROMATIC NUMBER [19].

Some of the results of this chapter for permutation graphs can be generalized to cocomparability graphs (i.e., complements of comparability graphs, Definition 3.1.1, page 28). We showed in section 7.5 that, if the treewidth of a cocomparability graph G is at most k, then the pathwidth is at most $O(k^2)$, and we gave a simple algorithm which finds a path-decomposition with this width. The results for permutation graphs presented in this chapter, i.e., the algorithm to compute the treewidth of a permutation graph and the proof that the treewidth and pathwidth are equal for permutation graphs, can be generalized to cocomparability graphs as follows.

1. The treewidth and pathwidth of *any* cocomparability graph are equal. Independently, this was also remarked in [79], and only very recently, R. H. Möhring reported to us that for graphs containing no astroidal triple, the treewidth and pathwidth are equal as well.

2. Using the results of [77], the treewidth can be computed in polynomial time for cocomparability graphs of which the partial order dimension of the complement is bounded by some constant.

3. In [79] it is shown that the results of this chapter can be extended to give an approximation for the treewidth of cocomparability graphs, of which the performance ratio depends on the partial order dimension of the complement.

Finally, we want to remark that the problem to determine the treewidth of cocomparability graphs in general, is NP-hard (this is also mentioned in [79]). In fact, it follows from [4], that finding the treewidth of complements of bipartite graphs is NP-hard.

9.1   Preliminaries

In this section we start with some definitions and easy lemmas. For more information on classes of perfect graphs the reader is referred to [15, 38, 76]. Recall the definition of an interval graph. There are many ways to characterize interval graphs. We restate the characterization of Lekkerkerker and Boland given in Lemma 2.1.8 (page 11).

Lemma 9.1.1 *An undirected graph is an interval graph if and only if the following two conditions are satisfied:*

1. *G is triangulated, and*

2. *every three vertices of G can be ordered in such a way that every path from the first to the third vertex passes through a neighbor of the second vertex.*

Three vertices which do not satisfy the second condition are called an *astroidal triple* (see Definition 2.1.7). These vertices are pairwise non-adjacent and for any pair of them there is a path that avoids the neighborhood of the remaining vertex. Let k be an integer. Recall that a graph G has pathwidth $\leq k$ if and only if there is a triangulation H of G such that H is an interval graph with $\omega(H) \leq k + 1$.

In this chapter we show that the treewidth and pathwidth of a permutation graph can be computed in polynomial time. Recall Definition 7.1.3 on page 78 of a permutation graph. In this chapter we assume that the permutation is given and we identify the permutation graph with the inversion graph. Recall that a permutation graph is an intersection graph, which is illustrated by the matching diagram (Figure 7.1, on page 79).

9.2 Separators obtained from scanlines

In this section we show that every minimal separator in a permutation graph can be obtained by using a *scanline*. Recall the definition of the matching diagram (Definition 7.1.4, page 79). (We call a matching diagram a diagram for short). It consists of two horizontal lines, one above the other, and a number of straight-line segments, one for each vertex, such that each line segment has one end vertex on each horizontal line. Two vertices are adjacent, if the corresponding line segments intersect. We say that two line segments *cross* if they have a nonempty intersection.

Definition 9.2.1 *A* scanline *in the diagram is any line segment with one end vertex on each horizontal line. A* scanline *s is* between *two non crossing line segments* x *and* y *if the top point of* s *is in the open interval bordered by the top points of* x *and* y *and the bottom point of* s *is in the open interval bordered by the bottom points of* x *and* y.

If a scanline s is between line segments x and y then the intersection of each pair of the three line segments is empty. Consider two nonadjacent vertices x and y. The line segments in the diagram corresponding to x and y do not cross in the diagram. Hence we can find a scanline s between the lines x and y. Take out all the lines that cross the scanline s. Clearly this corresponds with an x, y-separator in the graph. The next lemma shows that we can find all minimal x, y-separators in this way.

Theorem 9.2.1 *Let* G *be a permutation graph, and let* x *and* y *be non-adjacent vertices in* G. *Every minimal* x, y-*separator consists of all line segments crossing a scanline which lies between the line segments of* x *and* y.

Proof. Let S be a minimal x, y-separator. Consider the connected components of $G[V - S]$. Let C_x be the component containing x and C_y be the component containing y. Clearly these must also be 'connected' parts in the diagram, and we may assume without loss of generality that the component containing x is completely to the left of the component containing y. Every vertex of S is adjacent to some vertex in C_x and to some vertex in C_y (Lemma 2.1.3, page 9). Notice that we can choose a scanline s crossing no line segment of $G[V - S]$, and which is between x and y. Then all lines crossing the scanline must be elements of S. But for all elements of S the corresponding line segment must cross s, since it is intersecting with a line segment of C_x, which is to the left of s, and with a line segment of C_y, which is to the right of s.   □

Corollary 9.2.1 *There are $O(n^2)$ minimal separators in a permutation graph with n vertices.*

If s is a scanline, then we denote by S the set of vertices of which the corresponding line segments cross s. In the rest of this chapter we consider only scanlines of which the end points do not coincide with end points of other line segments.

Definition 9.2.2 *Two scanlines s_1 and s_2 are equivalent, denoted as $s_1 \equiv s_2$, if they have the same position in the diagram relative to every line segment; i.e. the set of line segments with the top (or bottom) end point to the left of the top (or bottom) end point of the scanline is the same for s_1 and s_2.*

Notice that $S_1 = S_2$ does not necessarily imply that $s_1 \equiv s_2$ (the converse implication of course holds).

We are only interested in scanlines which do not cross too many line segments, since these correspond with suitable separators.

Definition 9.2.3 *A scanline s is k-small if it crosses with at most $k + 1$ line segments.*

Lemma 9.2.1 *There are $O(nk)$ pairwise non-equivalent k-small scanlines.*

Proof. Consider the matching diagram, with numbers $1, \ldots, n$ written from left to right and underneath written $\pi_1, \ldots, \pi_n$. Consider a scanline t and assume the top end point is between i and $i + 1$ and the bottom end point is between π_j and π_{j+1}. Assume that α line segments are such that the top end point is to the left of the top of t and the bottom end point to the left of the bottom of t. Then the number of line segments crossing t is $i + j - 2\alpha$. Since $\alpha \leq i$ and $\alpha \leq j$, it follows that $i - k - 1 \leq j \leq i + k + 1$ must hold. This proves the lemma.   □

9.3 Treewidth and pathwidth of permutation graphs are equal

In this section we show that a permutation graph can be triangulated optimally such that the result is an interval graph. Recall the definition of a minimal triangulation of a graph (Definition 2.1.11, page 15).

Theorem 9.3.1 *Let* G *be a permutation graph, and let* H *be a minimal triangulation of* G*. Then* H *is an interval graph.*

Proof. Assume H has an astroidal triple x, y, z. Since x, y and z are pairwise nonadjacent, the corresponding line segments in the matching diagram pairwise do not cross. We may assume without loss of generality that the line segment of y is between those of x and z. Take a path p between x and z which avoids the neighborhood of y. Then each line of the path lies totally to the left or totally to the right of y. It follows that there are x' to the left of y and z' to the right of y such that x' and z' are adjacent in H, but neither x' or z' is a neighbor of y in H. Let S be a minimal x', y-separator in H. Since H is a minimal triangulation, S is also a minimal x', y-separator in G. By Lemma 9.2.1 S consists of all lines crossing some scanline s between x' and y. Clearly, the connected component of $G[V - S]$ containing x' lies totally to the left of s and the connected component containing z' in $G[V - S]$ lies totally to the right of s (notice $z' \notin S$ since z' lies totally to the right of y). It follows that x' and z' must be in different components of $G[V - S]$. Since H is minimal, by Theorem 2.1.1 they must also be in different components of $H[V - S]$. But then x' and z' can not be adjacent in H. It follows that there can not be an astroidal triple, and by the characterization of Lekkerkerker and Boland ([120], stated in Lemma 9.1.1), H is an interval graph.   □

Corollary 9.3.1 *For any permutation graph* G*, the pathwidth of* G *is equal to the treewidth of* G*.*

9.4 Candidate components

A graph with at least $k + 2$ vertices has treewidth at most k if and only if there is a minimal vertex separator with at most k vertices such that all components with S added as a clique have treewidth at most k. We can find a minimal separator using a scanline. The main result of this section is that we can change the diagram of a component in such a way that that the separator is changed into a clique.

Consider the matching diagram of G.

Definition 9.4.1 *Let s_1 and s_2 be two scanlines of which the intersection is either empty or one of the end points of s_1 and s_2. A candidate component $C = C(s_1, s_2)$ is a subgraph of G induced by the following sets of lines:*

- *all lines that are between the scanlines (in case the scanlines have a common end point, this set is empty), and*

- *all lines crossing at least one of the scanlines.*

We identify the candidate component $C = C(s_1, s_2)$ with the diagram containing s_1, s_2 and the set of lines corresponding with vertices of C.

Definition 9.4.2 *Let k be an integer. A candidate component $C = C(s_1, s_2)$ is k-feasible if there is a triangulation H of C such that $\omega(H) \leq k + 1$ and such that for each scanline s_i $(i = 1, 2)$ the set of lines crossing this scanline forms a clique in H.*

Notice that, if a candidate component has at most $k + 1$ vertices, then it is k-feasible.

Definition 9.4.3 *Let $C = C(s_1, s_2)$ be a candidate component. We define the realizer R(C) as the graph obtained from C by adding all edges between vertices of S_1 and between vertices of S_2 (i.e., the two subgraphs of R(C) induced by S_1 and by S_2 are cliques).*

A candidate component $C = C(s_1, s_2)$ is k-feasible if and only if the realizer R(C) has treewidth at most k.

Lemma 9.4.1 *If $C = C(s_1, s_2)$ is a candidate component, then the realizer R(C) is a permutation graph.*

Proof. Consider the matching diagram. Assume s_1 is to the left of s_2. First consider lines that cross only s_1 and with top end point to the right of the top end point of s_1. Let $(a_1, b_1), \ldots, (a_r, b_r)$ be these line segments with top end points $a_1, \ldots, a_r$. Assume $a_1 < a_2 < \ldots < a_r$. Change the order of $b_1, \ldots, b_r$ such that $b_1 > b_2 > \ldots > b_r$. This is illustrated in figure 9.1. Now consider the line segments crossing s_1 of which the top end point is to the left of the top end point of s_1. Reorder in the same way the *top end points* of these line segments. The lines crossing s_2 are handled similarly. The resulting diagram is a matching diagram for R(C).   □

Remark. Let $C = C(s_1, s_2)$ be a candidate component, such that the scanlines s_1 and s_2 have an end point in common. In this case the graph R(C) is chordal and therefore, there is an efficient algorithm to check whether the candidate component is k-feasible.

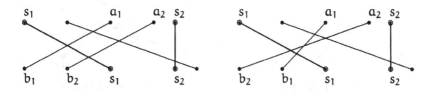

Figure 9.1: diagrams of candidate component and realizer

Let $C = \mathcal{C}(s_1, s_2)$ be a candidate component such that C has at least $k + 2$ vertices. The realizer $R(C)$ has treewidth at most k if and only if there is a minimal vertex separator S with at most k vertices such that all components, with S added as a clique, have treewidth at most k (see Lemma 2.1.15 on page 17). Consider the diagram of $R(C)$, obtained from the diagram of C by the method described in the proof of Lemma 9.4.1. By Lemma 9.2.1 a minimal separator can be found by a scanline. Let H be a minimal triangulation of $R(C)$ and let the scanline s represent a minimal vertex separator in H for non-adjacent vertices a and b. The separator consists exactly of the lines crossing this scanline s.

Definition 9.4.4 *Let* $C = \mathcal{C}(s_1, s_2)$ *be a candidate component with realizer* $R(C)$. *A scanline* t *is* nice *if the top point of* t *is in the closed interval between the top points of* s_1 *and* s_2, *and the bottom point of* t *is in the closed interval between the bottom points of* s_1 *and* s_2.

Lemma 9.4.2 *There is a scanline* $s^* \equiv s$ *such that* s^* *is nice.*

Proof. Consider the diagram of $R(C)$ with the scanlines s_1, s_2 and s. Without loss of generality, we assume s_1 is to the left of s_2. s separates non adjacent vertices a and b. Let the line segment of a be to the left of the line segment of b. The scanline s lies between the line segments of a and b. Assume s is not nice. Without loss of generality assume it crosses s_1. Then $a \in S_1$ and $b \notin S_1$ (since a and b are not adjacent). (see the left diagram in figure 9.2, page 112). Let s^* be the line segment with top point the top of s_1 and bottom point the bottom of s. We want to proof that $s \equiv s^*$. This clearly is the case if there is no line segment of which the top point is between the top points of s and s_1. Assume there is such a vertex p (see the right diagram in figure 9.2). Notice that, since p and a both cross s_1 they are adjacent. We claim that $S^* \subset S$. Let x be a line segment crossing s^*. If the bottom end of x is to the left of the bottom end of s^*, then the segment x clearly also crosses s. Assume the bottom vertex of x is to the right of the bottom vertex of s^*. Then the line segment also

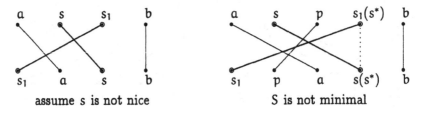

assume s is not nice   S is not minimal

Figure 9.2: there is an equivalent nice scanline

crosses s_1. But then x and a are adjacent, hence the top vertex of x must
be to the left of the top vertex of a. This implies that x also crosses s.
Clearly, S^* is an a, b-separator in $R(C)$. Since $p \in S \setminus S^*$, S cannot be a
minimal a, b-separator in $R(C)$, and since H is a minimal triangulation of
$R(C)$, it cannot be a minimal a, b-separator in H (Theorem 2.1.1). This is
a contradiction.   □

This proves that all lines crossing the scanline s are in C, and the next
lemma follows.

Lemma 9.4.3 *Let $C = C(s_1, s_2)$ be a candidate component with at least
$k+2$ vertices. Then C is k-feasible if and only if there is a nice scanline
s such that the two candidate components $C_1 = C(s_1, s)$ and $C_2 = C(s, s_2)$
are both smaller than C and are both k-feasible.*

Definition 9.4.5 *Two scanlines s_1 and s_2 are neighbors if they have one
endpoint in common and in the interval determined by the other end-
points, lies exactly one endpoint of a line segment.*

We are now ready to proof our main theorem.

Theorem 9.4.1 *Let $C = C(s_1, s_2)$ be a candidate components with s_2 to
the right of s_1. Then C is k-feasible if and only if there exists a sequence
of scanlines $s_1 = t_1, t_2, \ldots, t_r = s_2$ such that the following conditions are
satisfied:*

- *the scanlines t_i and t_{i+1} are neighbors for $i = 1, \ldots, r-1$, and one
  end point of t_{i+1} lies to the right of the end point of t_i.*

- *Each $C(t_i, t_{i+1})$ has at most $k+1$ vertices $(i = 1, \ldots, r-1)$.*

Proof. First assume such a sequence exists. Let $t_1, \ldots t_r$ be the sequence of
scanlines. If C has at most $k+1$ vertices, then C is k-feasible. Hence we
may assume that C has at least $k+2$ vertices. Then $r \geq 3$. By induction
we show that $C(t_1, t_i)$ is k-feasible for $i = 2, \ldots r$. If $i = 2$ then $C(t_1, t_2)$
has at most $k+1$ vertices and hence it is k-feasible. Assume $C(t_1, t_{i-1})$ is

k-feasible and $C(t_1, t_i) \neq C(t_1, t_{i-1})$. Then t_{i-1} is a nice scanline in $C(t_1, t_i)$. $C(t_1, t_{i-1})$ and $C(t_{i-1}, t_i)$ are both k-feasible, hence, by Lemma 9.4.3, also $C(t_1, t_i)$ is k-feasible.

Now assume that C is k-feasible. Consider the case that C has at most $k + 1$ vertices. In this case it is easy to see that such a sequence of scanlines exist. Assume C has at least $k + 2$ vertices. Then by Lemma 9.4.3 the is a nice scanline s^* such that $C(s_1, s^*)$ and $C(s^*, s_2)$ are both k-feasible. The theorem follows by induction on the number of vertices of C. □

9.5 An algorithm for the treewidth of a permutation graph

In this section we show how to compute the treewidth (and pathwidth) of a permutation graph G. Let k be an integer. The algorithm we present checks whether the treewidth of the permutation graph does not exceed k.

We use a directed acyclic graph $W_k(G)$ as follows. The vertices of the graph are the pairwise non-equivalent k-small scanlines. Direct an arc from scanline s to t if the following two conditions hold:

- The scanlines s and t are neighbors such that one end point of t is to the right of the correponding end point of s, and

- the candidate component $C(s, t)$ has at most $k + 1$ vertices.

We call this graph the *scanline graph*.

Lemma 9.5.1 *Let s_L be the scanline which lies totally to the left of all line segments and let s_R be the scanline which lies totally to the right of all line segments. G has treewidth at most k if and only if there is a directed path in the scanline graph from s_L to s_R.*

Proof. Clearly s_L and s_R are k-small. The result follows immediately from Theorem 9.4.1. □

Lemma 9.5.2 *The scanline graph has $O(nk)$ vertices and each vertex is incident with at most 4 arcs.*

Proof. The bound on the number of vertices is proved in Lemma 9.2.1. For each scanline s there are at most 4 scanlines t such that s and t are neighbors. □

The next three steps describe the algorithm which determines if the treewidth of G is at most k.

Step 1 Make a maximal list $\mathcal{L}$ of pairwise non-equivalent k-small scanlines.

Step 2 Construct the acyclic digraph $W_k(G)$.

Step 3 If there exists a path in $W_k(G)$ from s_L to s_R, then the pathwidth of G is at most k. If such a path does not exist, then the pathwidth of G is larger than k.

We discuss the running time of the algorithm in more detail.

Lemma 9.5.3 *The algorithm can be implemented to run in* $O(nk)$ *time.*

Proof. By Lemma 9.2.1 each k-small scanline can be characterized by two indices i and θ with $0 \le i \le n$ and $-(k+1) \le \theta \le k+1$. The scanline with these indices has top end point between i and $i+1$ and bottom end point between $\pi_{i+\theta}$ and $\pi_{i+\theta+1}$ (with obvious boundary restrictions). Let $A(i,\theta)$ be the number of line segments of which the top end point is to the left of the top endpoint of this scanline and which cross the scanline. Notice that $A(0,\theta) = 0$ for $\theta = 0, \ldots, k+1$. The rest of the table follows from:

$$A(i,\theta) \quad = \begin{cases} A(i,\theta-1) & \text{if } \pi_{i+\theta} > i \\ A(i,\theta-1) - 1 & \text{if } \pi_{i+\theta} \le i \end{cases}$$
$$\text{for } \theta > -\min(k+1,i), \text{ and}$$
$$A(i,\theta) = \begin{cases} A(i-1,\theta+1)+1 & \text{if } \pi_i^{-1} > i+\theta \\ A(i-1,\theta+1) & \text{if } \pi_i^{-1} \le i+\theta \end{cases}$$
$$\text{for } i \ge 1 \wedge \theta = -\min(k+1,i).$$

Notice that the number of line segments crossing the scanline with indices i and θ is $2A(i,\theta) + \theta$. It follows that the list $\mathcal{L}$ of k-small scanlines can be made in $O(nk)$ time.

We now show that the scanline graph $W_k(G)$ can be constructed in $O(nk)$ time. Consider two k-small scanlines s and t that are neighbors and which have a top end point in common, say between i and $i+1$. Assume the bottom end point of s is between π_{j-1} and π_j and the bottom end point of t is between π_j and π_{j+1}. Now notice that the number of vertices of $C(s,t)$ is $j - i + A(i,j-i) + A(i,j-i-1)$. This shows that the adjacency list for each k-small scanline can be computed in $O(k)$ time. Computing a path in W from s_L to s_R, if it exists, clearly also takes $O(nk)$ time. Hence the total algorithm can be implemented to run in $O(nk)$ time. □

Theorem 9.5.1 *Let G be a permutation graph. Then the pathwidth and treewidth of G are the same, and there is an* $O(nk)$ *algorithm that determines whether the treewidth of G is at most k.*

We end this section by noting that the algorithm can be adapted such that it computes, within the same time bound, the treewidth (resp. pathwidth) of a permutation graph (assuming that the matching diagram is given). This can be seen as follows. Let the treewidth of G be k. First compute a number L such that $L < k \leq 2L$. This can be done, using the algorithm described above $O(\log k)$ times, in time $O(nk)$ (execute the algorithm described above for $L = 1, 2, 4, \ldots$). In fact, this step can also be performed using the algorithm described in section 7.4. Now construct the scanline graph $W_{2L}(G)$, and put weights on the arcs, saying how many vertices are in the corresponding candidate component. Then search for a path from s_L to s_R, such that the maximum over weights of arcs in the path is minimized. This maximum weight minus one gives the exact treewidth k of G.

Chapter 10

Treewidth of circle graphs

Consider a set of n chords of a circle. Associate with this set an undirected graph as follows. The vertex set is the set of chords and two vertices are adjacent if and only if the corresponding chords intersect. Such a graph is called a *circle graph* and we call a set of chords representing the graph a *circle model*. In this chapter we do not distinguish between a circle graph and the circle model, i.e., we assume that we have a circle model of the graph. If the circle model is not given, it can be found in $O(nm)$ time [36, 37, 38, 67, 131].

Another recognition algorithm was presented in [51]. This characterization makes use of so-called *local complementations*. A local complementation of a graph and a vertex x is the operation which replaces the subgraph induced by the neighborhood of x by the complement. Two graphs are called locally equivalent if one can be obtained from the other by a successive series of local complementations. The characterization of circle graphs states the following. A graph G is a circle graph if and only if no graph locally equivalent to G has an induced 5-wheel, 7-wheel or $S(K_4)$ (i.e., the subdivision of K_4, see Definition 7.0.1, page 75 for the subdivision of a graph).

It is interesting to notice that the class of distance hereditary graphs is contained in the class of circle graphs. In [2] an $O((n + m)n^2)$ algorithm is given that computes the treewidth of a distance hereditary graph. Our algorithm computes the treewidth of circle graphs in $O(n^3)$ time.

Another interesting fact to note is that some important problems remain NP-complete when restricted to circle graphs. These problems include for example the CHROMATIC NUMBER problem [89], the COCHROMATIC NUMBER problem [73, 170] and the ACHROMATIC NUMBER problem [19]. There exists a heuristic for coloring circle graphs with performance guarantee of $O(\log n)$ [161]. On the other hand, some NP-complete problems are solvable in polynomial time, when restricted to circle graphs, for example the MAXIMUM INDEPENDENT SET, which can be solved in $O(n^2)$ time [160]. For more information on circle graphs and related classes of graphs we refer as usual to [38, 76].

In this chapter we give a simple and efficient algorithm to determine the treewidth of circle graphs. To illustrate the simplicity of the algorithm, and to whet the reader's appetite, we describe the algorithm here. Consider the circle model. Go around the circle in clockwise order and place a new vertex between every two consecutive end vertices of chords. Let Z be the set of these new vertices. Consider the polygon $\mathcal{P}$ with vertex set Z, and let T be a triangulation of this polygon. For a triangle in this triangulation define the weight as the number of chords in the circle model that cross the triangle. The weight of the triangulation T is the maximum weight of the triangles. The treewidth of the circle graph is the minimum weight minus one over all triangulations of the polygon $\mathcal{P}$. It is not hard to see that, using dynamic programming, the treewidth can be computed in $O(n^3)$ time.

Since the class of permutation graphs is properly contained in the class of circle graphs, our result generalizes some results of chapter 9 where an $O(nk)$ algorithm is given for the treewidth (and pathwidth) of permutation graphs, where k is the treewidth of the graph.

10.1 Preliminaries

Definition 10.1.1 *A* circle graph *is a graph for which one can associate for each vertex a chord of a circle such that two vertices are adjacent if and only if the corresponding chords have a nonempty intersection.*

Without loss of generality we can assume that no two chords share an end vertex. A set of chords of a circle such that the graph is isomorphic with the intersection graph is called a *circle model* for the graph. Throughout this chapter we identify a circle graph and a circle model of the graph, i.e., we assume that we have a circle model.

One of the main tools in this chapter is a method to locate all minimal vertex separators in circle graphs quickly.

10.2 Scanlines

As mentioned before, we assume that no two chords of the circle model share an end vertex.

Definition 10.2.1 *Place new points on the circle as follows. Go around the circle in clockwise order. Between every two consecutive end vertices of chords, place a new vertex. These new vertices are called* scanline *vertices.*

If n is the number of vertices in the circle graph then there are $2n$ scanline vertices. We denote the set of scanline vertices by Z.

Definition 10.2.2 *A* circle-scanline *is a chord of the circle of which the end vertices are scanline vertices.*

Definition 10.2.3 *Two circle-scanlines* cross *if they have a nonempty intersection but no end vertex in common.*

Definition 10.2.4 *Given two non crossing chords s_1 and s_2. A circle-scanline s is* between s_1 *and* s_2 *if every path from an end vertex of s_1 to an end vertex of s_2 along the circle passes through an end vertex of s.*

If a and b are nonadjacent vertices of the circle graph then the corresponding chords in the circle model do not cross. Take a circle-scanline s which is between the chords of a and b. Clearly, the set of vertices, corresponding to chords that cross s, is an a, b-separator. The following lemma is a generalization of Lemma 9.2.1 (page 107).

Lemma 10.2.1 *Let G be a circle graph and let a and b be non adjacent vertices. For every minimal a, b-separator S there exists a circle-scanline s between a and b such that the chords corresponding to vertices of S are exactly the chords crossing s.*

Proof. The proof is basically the same as the proof of Lemma 9.2.1.   □

Corollary 10.2.1 *There are at most $O(n^2)$ minimal vertex separators in a circle graph.*

10.3 Components and realizers

Let $G = (V, E)$ be a circle graph. Consider a circle model for G with the set Z of scanline vertices.

Definition 10.3.1 *Let $Y \subseteq Z$ be a set of scanline vertices with at least three elements. Consider the convex polygon $\mathcal{P}(Y)$ with vertex set Y. The* component $G(Y)$ *is the subgraph of G induced by the set of vertices corresponding with chords in the circle model which have a non empty intersection with the interior region of $\mathcal{P}(Y)$.*

Hence the edges of the polygon $\mathcal{P}(Y)$ are circle-scanlines. Notice that if $Y = Z$ then $G(Y)$ is simply the graph G.

Definition 10.3.2 *Let* Y *be a set of at least three scanline vertices and consider the component* $G(Y)$. *For each circle-scanline that is an edge of the polygon* $\mathcal{P}(Y)$, *add edges between vertices of* $G(Y)$ *of which the corresponding chords cross that circle-scanline. In this way we obtain the* realizer $R(Y)$ *of the component* $G(Y)$.

Hence each component is a subgraph of its realizer. We mention here that if Y is a set of three scanline vertices then

1. each chord corresponding with a vertex of $G(Y)$ intersects exactly two edges of $\mathcal{P}(Y)$ and hence

2. $R(Y)$ is a clique.

Lemma 10.3.1 *If* $G(Y)$ *is a component then the realizer* $R(Y)$ *is a circle graph.*

Proof. Let $s_1, s_2, \ldots, s_t$ be the circle-scanlines which are the edges of the polygon $\mathcal{P}(Y)$. Consider these circle-scanlines one by one.

Let $c_1, c_2, \ldots, c_\ell$ be the chords that cross some circle-scanline s_k. Call the end vertices of c_i, a_i and b_i. Choose a_i and b_i such that, when going along the chord from a_i to b_i first the circle-scanline s_k is crossed before the interior region of $\mathcal{P}(Y)$ is entered. Rearrange the order of the vertices $a_1, \ldots, a_\ell$ on the circle such that, afterwards, every pair of chords c_i and c_j cross. Notice that this only adds edges in the component between vertices with corresponding chords in $\{c_1, \ldots, c_\ell\}$.

In this way we obtain a circle model for the realizer. $\square$

We identify the realizer $R(Y)$ with a circle model for $R(Y)$ obtained as in the proof of Lemma 10.3.1.

Definition 10.3.3 *Let* $G(Y)$ *be a component with realizer* $R(Y)$. *A circle-scanline* s *in the circle model for* $R(Y)$ *is* Y-*nice if the end vertices of* s *are elements of* Y.

We now state one of our main results.

Lemma 10.3.2 *Let* $R(Y)$ *be a realizer of a component* $G(Y)$. *Let* a *and* b *be two non adjacent vertices in* $R(Y)$ *and let* S *be a minimal* a, b-*separator in* $R(Y)$. *Then there is a* Y-*nice circle-scanline* s *such that* S *consists of the vertices corresponding with the chords that cross* s.

Proof. Consider the circle model for $R(Y)$. Since a and b are not adjacent we know that there is a circle-scanline s (with end vertices in Z) between the chords of a and b such that the set of chords crossing s corresponds to S. Choose such a circle-scanline s with a minimum number of end vertices that are not in Y. If both end vertices of s are elements of Y then s is Y-nice. Assume this is not the case. Then s crosses with at least one circle-scanline s' which is an edge of the polygon $\mathcal{P}(Y)$. Let the end vertices of s be x and y chosen in such a way that, if we traverse s from x to y, then we first cross s' before entering the region of $\mathcal{P}(Y)$.

Let α and β be the end vertices of s' chosen such that α is on the same side of s as the chord a, and β is on the same side of s as the chord b.

Let s* be the circle-scanline with end vertices y and α and let s** be the circle-scanline with end vertices y and β.

Since a and b are non adjacent in R(Y) the corresponding chords of a and b do not both cross the circle-scanline s'. Assume that the chord of a does *not* cross with s'. We now consider two cases.

b and s' do not cross. Then s* and s** are both circle-scanlines between a and b. Let S* and S** be the corresponding separators. We claim that either $S^* \subseteq S$ or $S^{**} \subseteq S$. This can be seen as follows. Assume there is a chord p in the realizer crossing with s* but not with s and a chord q in the realizer crossing with s** but not with s. Then p and q must both cross with s'. But this is a contradiction since p and q cannot cross without also crossing s.

b and s' do cross. In this case s* is a circle-scanline between a and b. We claim that $S^* \subseteq S$. Assume there is a chord p in R(Y) that crosses with s* but not with s. Then p and b cannot cross. But this is a contradiction, since both p and b cross with s'.

This shows that either s* or s** is a circle-scanline between a and b of which the corresponding separator is a subset of the separator of s. But both s* and s** both have one more end vertex in Y. This proves the theorem.   □

Definition 10.3.4 *A component G(Y) is k-feasible if the realizer R(Y) has treewidth at most k.*

In other words, the component G(Y) is k-feasible if and only if there is a triangulation of the component such that each clique has at most $k+1$ vertices and such that for every circle-scanline which is an edge of $\mathcal{P}(Y)$ the set of vertices that cross that circle-scanline induce a clique in the triangulation.

Lemma 10.3.3 *Let Y be a set of scanline vertices with at least three elements such that the component G(Y) has at least $k+2$ vertices. Then the following two statements hold.*

1. *If G(Y) is k-feasible then there is a Y-nice circle-scanline, dividing the polygon $\mathcal{P}(Y)$ in two new polygons with vertex sets, say, Y_1 and Y_2, such that the components $G(Y_1)$ and $G(Y_2)$ both have less vertices than G(Y) and such that both $G(Y_1)$ and $G(Y_2)$ are k-feasible, and*

2. *if there is a Y-nice circle-scanline, dividing the polygon $\mathcal{P}(Y)$ in two new smaller polygons with vertex sets Y_1 and Y_2, such that the components $G(Y_1)$ and $G(Y_2)$ are both k-feasible, then $G(Y)$ is also k-feasible.*

Proof. Assume that $G(Y)$ is k-feasible. Consider a minimal triangulation H of $R(Y)$ (Definition 2.1.11, page 15). Since the number of vertices is at least $k + 2$ there must be a pair of non adjacent vertices a and b in H. Consider a minimal a, b-separator S in H. Since H is a minimal triangulation, S is also a minimal a, b-separator in G. Since H is chordal S induces a clique in H. By Lemma 10.3.2 there is a Y-nice circle-scanline s corresponding with S. s divides the polygon $\mathcal{P}(Y)$ into two new polygons. Let Y_1 and Y_2 be the vertex sets of these two new polygons.

We may assume that $G(Y_1)$ contains the chord corresponding with a and $G(Y_2)$ contains the chord corresponding with b. Let C_a and C_b be the vertex sets of $G(Y_1)$ and $G(Y_2)$ respectively. Then it follows that the induced subgraphs $H[C_a]$ and $H[C_b]$ are triangulations of $R(Y_1)$ and $R(Y_2)$, and hence $G(Y_1)$ and $G(Y_2)$ are k-feasible. Since a is not in $G(Y_2)$ and b not in $G(Y_1)$ it follows that both these components have less vertices than $G(Y)$.

Assume that there is a Y-nice circle-scanline s dividing the polygon $\mathcal{P}(Y)$ in two new smaller polygons with vertex sets Y_1 and Y_2, such that the components $G(Y_1)$ and $G(Y_2)$ are both k-feasible. Consider triangulations H_1 and H_2 of $R(Y_1)$ and $R(Y_2)$ respectively. Let S be the set of vertices corresponding with chords that cross s. Since S induces a clique in H_1 and in H_2, it follows that we can obtain a triangulation H of $R(Y)$ by identifying the vertices of S is H_1 and H_2. This shows that $G(Y)$ is k-feasible. □

Definition 10.3.5 *Let $\mathcal{P}$ be a polygon with m vertices. A triangulation of $\mathcal{P}$ is a set of $m - 3$ non crossing diagonals in $\mathcal{P}$ that divide the interior of $\mathcal{P}$ in $m - 2$ triangles.*

Definition 10.3.6 *Let Y be a set of at least three scanline vertices. Consider a triangulation T of $\mathcal{P}(Y)$. The weight of a triangle is the number of chords in the circle model that have a non empty intersection with the triangle. The weight of the triangulation, $w(T)$, is the maximum weight of a triangle.*

We can now state our main result.

Theorem 10.3.1 *Let Y be a set of at least three scanline vertices. A component $G(Y)$ is k-feasible if and only if there is a triangulation T with weight at most $k + 1$.*

Proof. First assume that $G(Y)$ is k-feasible. If $G(Y)$ has at most $k+1$ vertices, then *any* triangulation of $\mathcal{P}(Y)$ has weight at most $k+1$. We proceed with induction on the number of vertices of $G(Y)$. Assume $G(Y)$ has more than $k+1$ vertices. By the first part of Lemma 10.3.3 there is a Y-nice circle-scanline which divides the polygon $\mathcal{P}(Y)$ in two new polygons $\mathcal{P}(Y_1)$ and $\mathcal{P}(Y_2)$ such that the components $G(Y_1)$ and $G(Y_2)$ both have less vertices than $G(Y)$ and such that both $G(Y_1)$ and $G(Y_2)$ are k-feasible. By induction there are triangulations T_1 of $\mathcal{P}(Y_1)$ and T_2 of $\mathcal{P}(Y_2)$ both with weight at most $k+1$. Then, clearly, $T = T_1 \cup T_2$ is a triangulation of $\mathcal{P}(Y)$.

Now assume T is a triangulation of $\mathcal{P}(Y)$ with weight at most $k+1$. If Y has only three vertices, $G(Y)$ has at most $k+1$ vertices, and hence $G(Y)$ is k-feasible. We now proceed with induction on the number of vertices of Y. Take any diagonal of T. This divides the polygon into two smaller polygons with vertex sets Y_1 and Y_2 say. Since there are triangulations of $\mathcal{P}(Y_1)$ and of $\mathcal{P}(Y_2)$ with weight at most $k+1$ we may conclude that both $G(Y_1)$ and $G(Y_2)$ are k-feasible. But then by the second part of lemma 10.3.3 also $G(Y)$ is k-feasible.   □

10.4   Algorithm

In this section we describe an algorithm to find the treewidth of a circle graph.

Theorem 10.4.1 *Given a circle graph G with n vertices. There exists an $O(n^3)$ algorithm to determine the treewidth of G.*

Proof. First compute a circle model for G. As mentioned earlier this step can be performed in $O(n^2)$ time. We may assume that we can decide whether two chords cross in $O(1)$ time.

Clearly we may assume that $n > 1$. Determine a set of scanline vertices Z. Since $n > 1$, Z has at least four vertices, and hence the polygon $\mathcal{P}(Z)$ is well defined. The algorithm we describe finds a triangulation with minimal weight for $\mathcal{P}(Z)$.

First, for each of the circle-scanlines compute the number of chords that cross the circle-scanline. Since there are $O(n^2)$ circle-scanlines, and the test if a circle-scanline and a chord cross can be performed in $O(1)$ time, this step costs $O(n^3)$ time.

Use dynamic programming to find an optimal triangulation for $\mathcal{P}(Z)$. Let the scanline vertices be $s_0, s_2, \ldots, s_{\ell-1}$ ordered clockwise. Let $P(i, t)$ be the polygon defined by $s_i, \ldots, s_{i+t-1}$, where indices are to be taken modulo ℓ. We define $w(i, t)$ as the minimum weight of a triangulation of the polygon $P(i, t)$. Let $c(i, j)$ be the number of chords crossing the circle-scanline with end vertices s_i and s_j. Then $w(i,t)$ can be determined in $O(n^3)$ time using the following. Set all $w(i,2)$ equal to 0. For $t = 3, \ldots, \ell$, compute for all i:

$$w(i,t) \;=\; \min_{2 \leq j < t} \left(\max \left(w(i,j),\, w(i+j-1, t-j+1),\, F(i,j) \right) \right)$$

$$\text{where } F(i,j) \;=\; \frac{c(i, i+j-1) + c(i+j-1, i+t-1) + c(i, i+t-1)}{2}$$

Correctness follows from the fact that each chord crossing a triangle intersects exactly two sides of the triangle. The treewidth of G is $w(0, \ell) - 1$.   □

Finally, we show that it is also easy to find a triangulation of G with minimum clique number. Notice that the algorithm described above can be easily adapted to return a triangulation T of the polygon $\mathcal{P}(Z)$ with minimum weight. Define the graph $H(T)$ with the same vertex set as G as follows. Two vertices are adjacent in $H(T)$ if there is a triangle such that the chords corresponding with the vertices intersect this triangle. Notice that G is a subgraph of $H(T)$. We show that $H(T)$ has a perfect elimination scheme. Consider a vertex z of $\mathcal{P}(Z)$ which is *not* incident with a diagonal of T. This vertex is incident with exactly one triangle Q. Consider a vertex x of which the corresponding chord intersects Q but no other triangle. Then the neighborhood of x, $N(x)$, is a clique, hence x is simplicial. Notice also that the number of vertices in $N(x)$ is the weight of Q minus one, showing that the number of vertices in the clique $\{x\} \cup N(x)$ is equal to the weight of Q. Remove x from the graph $H(T)$ and the corresponding chord from the circle model. If there is no chord left in the circle model which intersects Q but no other triangle, then remove z from Z. Repeating this process gives a perfect elimination scheme, showing that $H(T)$ is chordal. This also shows that the number of vertices in a maximum clique of $H(T)$ is equal to the weight of the triangulation.

Chapter 11

Finding all minimal separators of a graph

Given a graph, one is often interested in finding subsets of vertices (or their cardinality) which possess a certain property. For example the CLIQUE NUMBER of a graph G is the maximum cardinality of a subset S such that G[S] is complete. Similar problems are the INDEPENDENCE NUMBER, the DOMINATION NUMBER or the CHROMATIC NUMBER. For many of these problems, it would be convenient if one could use a decomposition of the graph by means of certain separators.

In this chapter we give an algorithm which lists all minimal vertex separators in a graph. By now there is a large amount of literature concerned with listing all kinds of subgraphs in a graph, like cycle separators, cliques, spanning trees, (simple) cycles etc. (see for example [16, 41, 45, 44, 126, 138, 166]). For listing other kinds of combinatorial structures we refer to [75].

The usefulness of separators in graphs is perhaps best illustrated by the recent results for classes of graphs of bounded treewidth. For these classes, linear time algorithms exist for many NP-hard problems *exactly* because a decomposition can be made using *separators of bounded size* [3, 5, 7, 20, 99]. A decomposition of this type can be found in linear time (Chapter 15), however the huge constants involved in these algorithms do not make them of much practical use. Our results show that for many classes of graphs *efficient* decomposition algorithms exist, i.e., the size of the separators is of no importance.

Notice that, in general, the number of separators can be exponential, as the following example shows. Consider the graph consisting of two non adjacent vertices s and t, and a set of $\frac{n-2}{2}$ (internally) vertex disjoint paths of length 3 from s to t. The number of minimal s, t-separators in this graph is $2^{(n-2)/2}$.

In chapters 9, 10 and 12 however, it is shown that many important classes of graphs have a polynomial number of minimal vertex separators. These graph classes include permutation graphs, circular permutation graphs, trapezoid graphs, circle graphs, circular arc graphs, distance hereditary graphs, chordal bipartite graphs, cocomparability graphs of bounded dimension and weakly triangulated graphs.

A closely related, but somewhat different approach was surveyed in [162]. In this paper (see also [71]) it is shown that for many classes of graphs (for example chordal graphs, clique separable graphs and edge intersection graphs of paths in a tree or EPT-graphs) a decomposition by *clique separators* is possible, and it is illustrated that such a decomposition can also be used to solve efficiently many NP-complete problems like MINIMUM FILL-IN, MAXIMUM CLIQUE, GRAPH COLORING and MAXIMUM INDEPENDENT SET. In [173] an algorithm is given for finding clique separators efficiently (the algorithm uses $O(nm)$ time to find one clique separator). Our results generalize the above mentioned results in the sense that at least some of these NP-hard problems are solvable for many other graph classes also, i.e., graph classes for which the number of minimal separators is polynomial bounded (Chapters 8, 9, 10 and 12, and [22, 55, 97, 113]).

In [91] an algorithm is given which finds all, what the authors call *minimum size separators*. By this they mean that given a graph which is k-connected, the algorithm finds all separators with k vertices. Moreover, they show in this paper that the number of these separators is bounded by $O(2^k \frac{n^2}{k})$. Their algorithm which lists all minimum size separators runs in time $O(2^k n^3)$.

There is another closely related concept, which we call *inclusion minimal separators*. An inclusion minimal separator is a minimal separator with the additional constraint that no proper subset is also a minimal separator. Notice that inclusion minimal separators lie more or less between the minimum size separators and the minimal separators, i.e., all minimum size separators are inclusion minimal and all inclusion minimal separators are minimal separators.

A subset S of vertices is a minimal separator of a graph $G = (V, E)$ if and only if there are *two* connected components in $G[V \setminus S]$ such that every vertex of S has at least one neighbor in each of these components (Lemma 11.1.1). A similar characterization can be given for inclusion minimal separators. A separator S of a graph $G = (V, E)$ is inclusion minimal if and only if every vertex of S has a neighbor in *every* connected component of $G[V - S]$. It follows that a list of all inclusion minimal separators can easily be obtained from the list of all minimal separators. However, until now we have not been able to find an *efficient* algorithm which finds all inclusion minimal separators.

The following example shows that the minimum size separators and the inclusion minimal separators are only of limited use. Consider any graph G. Take a new vertex x and make this adjacent to all vertices of G. Take another new vertex y and make this adjacent to x. Call this new graph H. The only inclusion minimal separator which is also the only minimum size separator of H is $\{x\}$. However if S is some minimal separator of G, then $S \cup \{x\}$ is a minimal separator in H.

11.1 Preliminaries

The following lemma is an simple extension of Lemma 2.1.3, page 9. It provides an easy test whether a given set S of vertices is a minimal separator or not.

Lemma 11.1.1 *Let S be a separator of the graph $G = (V, E)$. Then S is a minimal separator if and only if there are two different connected components of $G[V - S]$ such that every vertex of S has a neighbor in both of these components.*

Proof. Let S be a minimal a, b-separator and let C_a and C_b be the connected components containing a and b respectively. Let $x \in S$. Since S is a *minimal* a, b-separator, there is a path between a and b passing through x but using no other vertex in S. Hence x must have a neighbor in C_a and in C_b.

Now let S be a separator and let C_a and C_b be two connected components such that every vertex of S has a neighbor in C_a and in C_b. Let $a \in C_a$ and $b \in C_b$. Then clearly S is a minimal a, b-separator, for if $x \in S$, then there is a path between a and b which uses no vertices of $S \setminus \{x\}$.   □

Notice that this also proves the following. Let S be a minimal separator and let C_1 and C_2 be two connected components of $G[V - S]$ such that every vertex of S has a neighbor in both C_1 and C_2. If a is a vertex of C_1 and b is a vertex of C_2 then S is a minimal a, b-separator.

It may be a bit surprising at first sight that it is very well possible for one minimal separator to be contained in another one. This can easily be seen as follows. Consider a graph G. Take a new vertex x and make this adjacent to all vertices of G. Take another new vertex y and make this adjacent to x only. Let H be the new graph. Then, in H, all minimal separators contain x, and $\{x\}$ itself is also a minimal separator. Another example of this can be found in [76].

For minimal a, b-separators, however things are different, since by definition one minimal a, b-separator cannot be contained in another one.

We now show that at least some of the minimal separators are easy to find.

Definition 11.1.1 *Let a and b be non adjacent vertices. If S is a minimal a, b-separator which contains only neighbors of a then S is called close to a.*

Lemma 11.1.2 *If a and b are non adjacent then there exists exactly one minimal a, b-separator close to a.*

Proof. Let S be a minimal a, b-separator close to a. For every vertex in S there is a path to b which does not use any other neighbors of a, since S is minimal. On the other hand, if x is a neighbor of a such that there is a path to b without any other neighbors of a, then x must be an element of S, otherwise there is a path between x and b which avoids S and this is a contradiction since x is in the component of $G[V - S]$ that contains a. □

Notice that a minimal separator close to a can easily be computed as follows. Start with $S = N(a)$. Clearly, since a and b are non adjacent S separates a and b. Let C_b be the connected component of $G[V - S]$ containing b. Let $S' \subseteq S$ be the set of those vertices of S which have at least one neighbor in C_b. By Lemma 11.1.1 S' is a minimal a, b-separator, and since it only contains neighbors of a, it is close to a.

Lemma 11.1.3 *Let S be a minimal a, b-separator close to a and let C_a and C_b be the connected components containing a and b respectively. Let $S^* \neq S$ be another minimal a, b-separator. Then $S^* \subset S \cup C_b$.*

Proof. Since S^* is a minimal a, b-separator $S^* \subset C_a \cup C_b \cup S$. Assume S^* has a vertex $x \in C_a$. $S^* \setminus \{x\}$ does not separate a and b hence there is a path P between a and b using x but no other vertex of S^*. Since S is a minimal separator, P goes through a vertex $y \in S$. Since S is close to a, y is adjacent to a. Hence there is a path $P' \subset P$ between a and b that does not contain x. This is a contradiction since P' contains no vertex of S^*. □

In the next two sections we show how to obtain new minimal a, b-separators from a given one using so called minimal pairs. A minimal pair is in some sense the smallest step to go from one minimal a, b-separator to the next one. The main difficulty is to prove that we indeed obtain all minimal separators by using small steps only.

In section 11.4 we describe an algorithm that computes all minimal a, b-separators for a given pair of non adjacent vertices a and b in a breadth-first-search manner, we prove that it is correct and we analyse its time complexity. We end with some concluding remarks and some open problems.

11.2 Good pairs

Let $G = (V, E)$ be a graph and let a and b be non adjacent vertices in G. Let S be a minimal a, b-separator and let C_a and C_b be the connected components containing a and b respectively.

Definition 11.2.1 *Let* $\Delta \subseteq C_a \setminus \{a\}$ *and let* C'_a *be the connected component of* $G[C_a - \Delta]$ *that contains* a. *Let* $N \subseteq S$ *be the set of vertices in* S *that do not have a neighbor in* C'_a. *The pair* (Δ, N) *is called* good *for* S *if the following conditions are satisfied.*

1. $N \neq \emptyset$.

2. *Each* $\delta \in \Delta$ *has at least one neighbor in* C'_a.

3. *Each* $\delta \in \Delta$ *either has a neighbor in* N *or there exists a vertex* $x \in N$ *and a connected component* D *of* $G[C_a - \Delta]$ *such that both* x *and* δ *have at least one neighbor in* D.

Lemma 11.2.1 *If* S *is close to* a *then there is no good pair.*

Proof. Assume (Δ, N) is a good pair. Hence $\Delta \subseteq C_a \setminus \{a\}$. Let C'_a be the connected component of $G[C_a - \Delta]$ that contains a. The set N is defined as the subset of S that does not contain any neighbor in C'_a. Then $N = \emptyset$ since S contains only neighbors of a. But by definition $N \neq \emptyset$. $\square$

Theorem 11.2.1 shows that a good pair defines a new separator. In Theorem 11.2.2 we show that each minimal a, b-separator can be obtained by a good pair for the separator that is close to b. In section 11.3 we show that only a restricted type of good pairs, called minimal pairs, have to be considered.

Theorem 11.2.1 *Let* (Δ, N) *be a good pair for* S. *Define* $S^* = (S \cup \Delta) \setminus N$. *Then* S^* *is a minimal* a, b-*separator.*

Proof. Let C'_a be the connected component of $G[C_a - \Delta]$ that contains a. Clearly, S^* separates a and b, since vertices of N do not have neighbors in C'_a. Let C'_b be the connected component of $G[V - S^*]$ that contains b. Notice that $C_b \subset C'_b$, and since each vertex of N has a neighbor in C_b, $N \subset C'_b$.

Each vertex of S^* has at least one neighbor in C'_a by definition of a good pair, and each vertex of $S^* \setminus \Delta$ has at least one neighbor in C'_b since it has at least one neighbor in C_b. The only thing left to show is that each vertex of Δ has a neighbor in C'_b. Let $\delta \in \Delta$. By definition, either δ has a neighbor in N (and hence in C'_b) or there is a vertex $x \in N$ and a connected component D of $G[C_a - \Delta]$ such that both δ and x have a neighbor in D. D is also connected in $G[V - S^*]$ and since x has a neighbor in D, $D \subset C'_b$. $\square$

Theorem 11.2.2 *Assume S is close to b. Let $S^* \neq S$ be a minimal a, b-separator. There exists a good pair (Δ, N) such that $S^* = (S \cup \Delta) \setminus N$.*

Proof. Let C_a^* and C_b^* be the connected components of $G[V - S^*]$ containing a and b respectively.

First notice that $S^* \subset C_a \cup C_b \cup S$, since S^* is minimal. Since S is close to b, by Lemma 11.1.3, $S^* \subset S \cup C_a$. Let $\Delta = S^* \cap C_a$ and $N = S \setminus S^*$. We show that (Δ, N) is a good pair.

Since $S^* \neq S$ and both are minimal a, b-separators: $N \neq \emptyset$.

Let C_a' be the connected component of $G[C_a - \Delta]$ containing a. We show that N is exactly the set of vertices in S which do not have a neighbor in C_a'. In order to do this we claim that $C_a' = C_a^*$. Since C_a' is a connected component of $G[V - (\Delta \cup S)]$ and since $S^* \subset \Delta \cup S$, $C_a' \subseteq C_a^*$. Now assume there is a vertex $x \in N$ which has a neighbor $y \in C_a'$. Since S is close to b, x is a neighbor of b. This is a contradiction since there would be a path between a and b which does not use any vertex of S^*. This shows that $C_a' = C_a^*$. Since S^* is minimal, N is exactly the set of vertices in S that do not have a neighbor in C_a', and every vertex of $\Delta \cup (S \setminus N)$ has at least one neighbor in C_a'.

To prove the last item first notice that $N \subset C_b^*$ and that C_b^* contains exactly those connected components D of $G[C_a - \Delta]$ for which there is a vertex $y \in N$ which has a neighbor in D. Now let $\delta \in \Delta$. Since S_a^* is minimal, δ has a neighbor x in C_b^*. Since δ only has neighbors in $C_a \cup S$, x must be an element of N or of some component D of $G[C_a - \Delta]$. In this second case, there must also be a vertex $y \in N$ which has a neighbor in D.

This proves the theorem.   □

11.3   Minimal pairs

Again let $G = (V, E)$ be a graph and let a and b be non adjacent vertices in G. Let S be a minimal a, b-separator and let C_a and C_b be the connected components of $G[V - S]$ containing a and b respectively. In this section we show how to find some good pairs.

Definition 11.3.1 *Let $x \in S$ be non adjacent to a. Let $C_a(x)$ be the subgraph induced by $C_a \cup \{x\}$. Let Δ be the minimal x, a-separator in $C_a(x)$ close to x, and let C_a' be the connected component containing a. Now let N be the set of vertices of S which do not have a neighbor in C_a'. The pair (Δ, N) is called the minimal pair for S and x.*

Lemma 11.3.1 *A minimal pair is good.*

Proof. Notice that $x \in N$, hence $N \neq \emptyset$.

Now, Δ is a minimal x, a-separator in $C_a(x)$ and hence every vertex of Δ has a neighbor in C'_a.

Finally, if $\delta \in \Delta$ then δ is adjacent to x since Δ is close to x. Hence each vertex of Δ has a neighbor in N. $\qquad\square$

We want to prove that we can find every minimal a, b-separator by starting with the minimal a, b-separator that is close to b and by recursively using minimal pairs. The following technical lemma proves this.

Lemma 11.3.2 *Let (Δ, N) be a good pair for S. Let $x \in N$ and let (Δ^*, N^*) be the minimal pair for S and x. Let $S^* = (S \cup \Delta^*) \setminus N^*$. Define $\overline{\Delta} = \Delta \setminus \Delta^*$ and $\overline{N} = (N \setminus N^*) \cup (\Delta^* \setminus \Delta)$. Then:*

1. *if $\overline{N} = \emptyset$ then $(S \cup \Delta) \setminus N = S^*$, and if*

2. *$\overline{N} \neq \emptyset$ then $(\overline{\Delta}, \overline{N})$ is a good pair for S^* and $(S \cup \Delta) \setminus N = (S^* \cup \overline{\Delta}) \setminus \overline{N}$.*

Proof. We start with some easy observations. Let C'_a be the connected component of $G[C_a - \Delta]$ that contains a and let C^*_a be the connected component of $G[C_a - \Delta^*]$ that contains a. Let $\Delta' = N(x) \cap \Delta$.

- $C'_a \subseteq C^*_a$ since Δ^* contains no vertices of C'_a.

- $\Delta' \subseteq \Delta^*$ since every vertex of Δ' has a neighbor in C'_a.

- $\Delta \setminus \Delta' \subseteq C^*_a$ since every vertex of Δ has a neighbor in C'_a.

- $N^* \subseteq N$, since $C'_a \subseteq C^*_a$.

- C'_a is exactly the connected component of $G[C^*_a - (\Delta \setminus \Delta')]$ containing a since $C^*_a - (\Delta \setminus \Delta')$ contains all vertices of C'_a but no vertex of Δ.

- The set of vertices in S^* without a neighbor in C'_a is exactly $\overline{N}$, which is easy to check.

Assume $\overline{N} = \emptyset$. Then $\Delta^* \subseteq \Delta$ and $N = N^*$ (since $N^* \subseteq N$). Now clearly, also $\Delta^* = \Delta$, otherwise S^* and $(S \cup \Delta) \setminus N$ are two minimal a, b-separators of which one is properly contained in the other which is impossible by definition. Hence $S^* = (S \cup \Delta) \setminus N$.

Now assume $\overline{N} \neq \emptyset$. We show that $(\overline{\Delta}, \overline{N})$ is good for S^*. Notice that every vertex of $\overline{\Delta}$ has a neighbor in C'_a, since this holds for every vertex of Δ.

Let $\delta \in \overline{\Delta}$ and assume that δ has no neighbors in $\overline{N}$. Since $\delta \in C^*_a$, δ has no neighbor in N^*. Hence δ has no neighbor in N. Now (Δ, N) is a good pair, hence there is a vertex $z \in N$ and a connected component D of $G[C_a - \Delta]$ such that δ and z have a neighbor in D. Assume by way of contradiction that for no vertex of $\overline{N}$ there is a connected component in $G[C_a^* - \overline{\Delta}]$ such that this vertex and δ both have a neighbor in this component. The following observations lead to a contradiction.

- $N(\delta) \cap D \subseteq C_a^*$. Otherwise, since $\Delta^* \setminus \Delta' \subset \overline{N}$, δ has a neighbor in $\overline{N}$.

- $G[D \setminus \Delta^*]$ is connected. Since otherwise every connected component has a vertex with a neighbor in $\Delta^* \setminus \Delta$, and hence there is a connected component and some vertex in $\overline{N}$ such that this vertex and δ both have a neighbor in this component.

- D contains no vertices of Δ^*, by the same argument.

This shows that $D \subset C_a^*$. If $z \in N^*$ then z can have no neighbors in D, since z has no neighbors in C_a^*. Hence $z \in N \setminus N^*$. This is a contradiction, since now there is a connected component D in $G[C_a^* - \overline{\Delta}]$ and a vertex $z \in \overline{N}$ such that z and δ both have a neighbor in D.

The fact that $(S \cup \Delta) \setminus N = (S^* \cup \overline{\Delta}) \setminus \overline{N}$ is obvious. $\square$

By Lemma 11.3.2 and Theorem 11.2.2 we obtain the following result.

Corollary 11.3.1 Let S be a minimal a, b-separator and let S_1 be the minimal a, b-separator close to b. There exists a sequence $(\Delta_1, N_1), \ldots, (\Delta_t, N_t)$ such that

1. (Δ_1, N_1) is a minimal pair for S_1 and some vertex $x_1 \in N_1$.

2. For $i = 2, \ldots, t$, (Δ_i, N_i) is a minimal pair for $S_i = (S_{i-1} \cup \Delta_{i-1}) \setminus N_{i-1}$ and some vertex $x_i \in N_i$.

3. For $i = 1, \ldots, t$, Δ_i and a are in the same connected component of $G[V - S_i]$.

4. $S = (S_t \cup \Delta_t) \setminus N_t$.

11.4 An algorithm which finds all minimal separators

In this section we give an algorithm that, given a graph G and two non adjacent vertices a and b finds all minimal a, b-separators. This algorithm is displayed in figure 11.1 (page 133).

Theorem 11.4.1 Let S be the minimal a, b-separator that is close to b and let $\mathcal{T} = \{S\}$ and $\mathcal{Q} = \{S\}$. Then a call separators$(G, a, b, \mathcal{T}, \mathcal{Q})$ determines a set $\mathcal{Q}$ containing all minimal a, b-separators.

procedure separators$(G, a, b, \mathcal{T}, \mathcal{Q})$
input: Graph G and non adjacent vertices a and b and
    sets $\mathcal{T}$ and $\mathcal{S}$ of minimal a, b-separators.
output: Set $\mathcal{Q}$ of all minimal a, b-separators in G.
begin
    $\mathcal{T}' := \emptyset$;
    for each $S \in \mathcal{T}$ **do**
    begin
        Determine C_a;
        { C_a is the connected component of $G[V - S]$ that contains a.}
        for each $x \in S$ which is not adjacent to a **do**
        begin
            Determine Δ;
            {Δ is the minimal x, a-separator in $C_a(x)$ that is close to x.}
            Determine C'_a;
            { C'_a is the connected component of $G[C_a - \Delta]$ that contains a.}
            Determine N;
            {N is the set of vertices in S that do not have a neighbor in C'_a.}
            $S^* := (S \cup \Delta) \setminus N$;
            $\mathcal{T}' := \mathcal{T}' \cup \{S^*\}$
        end for
    end for;
    $\mathcal{T}' := \mathcal{T}' \setminus \mathcal{Q}$; $\mathcal{Q} := \mathcal{Q} \cup \mathcal{T}'$;
    separators$(G, a, b, \mathcal{T}', \mathcal{Q})$
end.

Figure 11.1: Algorithm finding minimal separators

Proof. By Corollary 11.3.1 the set Q contains all minimal a, b-separators. By Lemma 11.3.1 and Theorem 11.2.1 all sets in Q are minimal separators.
$\square$

Remark. If we let $T = \{\{b\}\}$ and $Q = \emptyset$ then a call separators(G, a, b, T, Q) has the same result.

Theorem 11.4.2 *Let* R *be the number of minimal* a, b-*separators (for non adjacent vertices* a *and* b *). The algorithm to determine all minimal* a, b-*separators can be implemented to run in time* $O(n^3 R)$.

Proof. Assume that the graph is given with an adjacency matrix. The minimal separator S that is close to b can easily be found in $O(n^2)$ time as follows. Initialize $S = N(b)$. Determine the connected component C_a of $G[V - S]$. Remove vertices from S that do not have a neighbor in C_a.

Consider the time it takes to compute T'. For each $S \in T$ and for each $x \in S$ not adjacent to a we have to do the following computations. Determining Δ again takes at most $O(n^2)$ time. Also computing C_a' and N can clearly be done in $O(n^2)$ time. Hence the time to compute T' (which may contain elements which are already in Q) can be performed in $O(n^3 | T |)$ time.

We have to remove minimal separators from the new set T' which have been found before. We can do this by keeping Q in a suitable data structure, allowing an update in $O(n| T' | \log R) = O(n^2 | T' |) = O(n^3 | T |)$ time. It follows that the computation of T', containing only new minimal separators, can be performed in $O(n^3 | T |)$ time.

Since each newly computed set T' contains only minimal separators that have not been found before, it follows that the total time needed by the algorithm is $O(n^3 R)$.
$\square$

Corollary 11.4.1 *The set of all minimal separators of a graph can be found in* $O(n^5 R)$ *time, where* n *is the number of vertices in the graph and* R *is the total number of minimal separators.*

A somewhat different results is the following.

Theorem 11.4.3 *Assume* G *has at least* R^* *minimal separators. There exists an algorithm which finds* R^* *different minimal separators in* $O(n^5 R^*)$ *time.*

Proof. The claimed algorithm is simply the following. The algorithm finding minimal separators described above is stopped when R^* different ones have been found. It may take time at most $O(n^5 R)$, trying different pairs of non-adjacent vertices for which the total number of different minimal separators is smaller than R. Assume a pair of vertices has been found with enough new minimal separators. Now the analysis in the proof of Theorem 11.4.2 shows the claimed result.   □

Chapter 12

Treewidth and pathwidth of cocomparability graphs of bounded dimension

This paper gives a polynomial time algorithm for solving the treewidth and pathwidth problems on cocomparability graphs of any fixed dimension d.

The dimension of a partial order is one of the most carefully studied parameters of a partial order [165]. Yannakakis [174] showed that determining whether a partial order has dimension at most d is NP-complete for any fixed $d \geq 3$. Many problems have been shown to be efficiently solvable on partial orders of dimension 2. However no NP-complete partial order problem was known that is solvable by a polynomial time algorithm for partially ordered sets of some fixed dimension greater than 2.

In chapter 9 we have shown that treewidth and pathwidth can be solved in polynomial time for cocomparability graphs of fixed dimension d assuming that a d-dimensional representation is given as input. If the graph is given as input, this does not yield a polynomial time algorithm, since we cannot find the representation efficiently. Thus, it was unclear whether the problem was easy because the class of graphs is well behaved, or because having the representation (which is the solution to an NP-complete problem) is such a powerful tool that it gave the solution.

We show that treewidth and pathwidth are solvable on cocomparability graphs of fixed dimension in polynomial time, when the graph is given as input. This shows that small dimension is capturing a property which makes the problems simpler. Note that by Yannakakis' result [174], we are not able to run a recognition algorithm to determine whether our input is in the class of graphs we study here. However, this algorithm is designed so that if the input graph G is not a cocomparability graph the algorithm reports this, and otherwise it solves the treewidth and pathwidth problem for G correctly or reports that G has dimension larger than d. Thus, the result of the algorithm can be used reliably even if we do not know *a priori* that G is in the class.

If the input G is not a cocomparability graph of dimension d, there are three possible outcomes. If G is not a cocomparability graph, which can be checked in time $O(n^{2.37\cdots})$ [155], we reject the input. If G is a cocomparability graph, we can use the algorithm given in Theorem 11.4.3. (page 134) to either decide that the number of minimal separators is at larger than n^d, or listing all minimal separators. This algorithm can be implemented to run in time $O(n^{d+5})$. If the number of minimal separators is larger than n^d, we reject the input. If the number of minimal separators is at most n^d and G is a cocomparability graph, the algorithm computes the treewidth and pathwidth correctly, even if G has dimension greater than d.

This turns the TREEWIDTH and the PATHWIDTH problem into the first known problems for which restricted dimension helps, while both problems are NP-complete for cocomparability graphs when the dimension is unbounded [4].

12.1 Preliminaries

Recall Lemma 11.1.1, page 127: A subset S is a minimal separator if and only if there are two connected components such that every vertex of S has a neeighbor in each.

The following lemma describes a useful property of minimal separators.

Lemma 12.1.1 *Let* $G = (V, E)$ *be a graph and* S *a minimal separator and a clique of* G. *Let* C *be a connected component of* $G[V \setminus S]$ *and let* x *and* y *be non adjacent vertices of* $S \cup C$. *Then every minimal* x, y*-separator* S^* *of* G *is a proper subset of* $S \cup C$.

Proof. Let S^* be a minimal x, y-separator of G and let C_x and C_y be the components of $G[V \setminus S^*]$ containing x and y, respectively.

C_x and C_y can not both have non-empty intersection with S since S is a clique. W.l.o.g. let $C_x \cap S = \emptyset$. Since S is a minimal separator of G, C_x is connected and $x \in C$ belongs to C_x, we get $C_x \subseteq C$. A vertex $s \in V \setminus (S \cup C)$ belongs to a component of $G[V \setminus S]$ different from C and can not have a neighbour in C_x. Hence, $s \notin S^*$. Finally, $S^* \subset S \cup C$ since $x, y \notin S^*$. □

One of the main tools in our algorithm is the fact that all minimal separators of a graph can be computed in time polynomial times the number of minimal separators. Let R be the number of minimal separators of a graph G. Then there exists an algorithm which list all minimal separators in G in time $O(n^5 R)$. This was shown in chapter 11.

We shall be trying to find an efficient triangulation of the graph (Definition 2.1.12, page 16). Recall Theorem 2.1.2: for every triangulation H of the graph, there exists an efficient triangulation which is a subgraph of H.

The treewidth problem is NP-hard even when restricted to bipartite or to cobipartite graphs [4]. Consequently, the TREEWIDTH problem is NP-hard when restricted to cocomparability graphs, which contain the cobipartite graphs as a proper subclass. Thus, a bound on the dimension is necessary (unless P=NP) and quite natural to enable the design of a polynomial time treewidth algorithm for cocomparability graphs.

Recall Definition 3.1.1 (page 28) of a comparability graph.

Definition 12.1.1 *A cocomparability graph is a graph of which the complement is a comparability graph.*

Permutation graphs (Definition 7.1.3) can be characterized as being exactly the graphs which are comparability and cocomparability graphs and they are exactly the comparability graphs of poset dimension at most two.

In [77] it is shown that cocomparability graphs are the intersection graphs of a concatenation of matching diagrams (Definition 7.1.4, page 79) (see also Lemma 7.1.1, page 78; the result stated there is somewhat weaker). The minimal number of matching diagrams needed, plus one is called the dimension of the cocomparability graph (in fact, this is equal to the dimension of the poset corresponding to the complement). Notice that a permutation graph is a cocomparability graph of dimension two. The following lemma is an immediate generalization of Lemma 9.2.1 (page 107) and Corollary 9.2.1.

Lemma 12.1.2 *A cocomparability graph of dimension d has at most n^d minimal separators.*

Recall the characterization of interval graphs discovered by Gilmore and Hoffman (Theorem 2.1.6, page 11): A graph is an interval graph if and only if there is an interval ordering of the maximal cliques. Using this characterization, we can easily identify the minimal vertex separators in an interval graph.

Lemma 12.1.3 *Let G be an interval graph and let $C_1, C_2, \ldots, C_t$ be an interval ordering of the maximal cliques of G. The minimal separators of G are the sets $C_i \cap C_{i+1}$, $(i = 1, \ldots, t - 1)$.*

Proof. Since each C_i is a maximal clique, we have that for each $1 \le i < t$: $C_i \setminus C_{i+1} \ne \emptyset$ and $C_{i+1} \setminus C_i \ne \emptyset$. Let $x \in C_i \setminus C_{i+1}$ and $y \in C_{i+1} \setminus C_i$. Then clearly $C_i \cap C_{i+1}$ is a minimal x, y-separator.

Now consider nonadjacent vertices a and b and let S be a minimal a, b-separator. Assume a appears before b in the clique arrangement. Let C_i be the last clique that contains a and let C_j be the first clique that contains

b. If S contains not all vertices of $C_i \cap C_{i+1}$, then there is a path from a to all vertices of $C_{i+1} \setminus S$ without using vertices of S.

Continuing in this way we either find a path from a to b or some $i \leq k < j$ such that $C_k \cap C_{k+1} \subseteq S$. □

The pathwidth problem is concerned with finding a triangulation of a graph into an interval graph such that the clique number is minimized. In general the pathwidth of a graph is at least equal to the treewidth of the graph. Determining the pathwidth of a graph is NP-complete, even when restricted to chordal graphs [78]. However, for cocomparabiltity graphs the measures treewidth and pathwidth coincide (chapter 9). We reformulate Theorem 9.3.1, page 109, for cocomparability graphs.

Theorem 12.1.1 *For a cocomparability the pathwidth and treewidth are equal. Moreover, every minimal triangulation (and hence also every efficient triangulation) is an interval graph.*

12.2 Pieces

In this section we assume that $G = (V, E)$ is a connected cocomparability graph with n vertices and of dimension d.

Definition 12.2.1 *Two minimal separators S_1 and S_2 are non-crossing if all vertices of $S_1 \setminus S_2$ are contained in the same connected component of $G[V \setminus S_2]$ and all vertices of $S_2 \setminus S_1$ are contained in the same connected component of $G[V \setminus S_1]$.*

Lemma 12.2.1 *Let H be a chordal graph. Then every pair of minimal separators in H is non-crossing.*

Proof. Let S_1 and S_2 be minimal separators. Since the graph is chordal, the subgraphs induced by these separators are cliques. It follows that $S_1 \setminus S_2$ is contained in one connected component of $H[V \setminus S_2]$. □

Lemma 12.2.2 *Let G_c be an efficient triangulation of G and let S_1, S_2 be minimal separators in G_c. Then S_1 and S_2 are non-crossing separators in G.*

Proof. S_1 and S_2 are non-crossing in G_c by Lemma 12.2.1. Since G_c is efficient, S_1 and S_2 are minimal separators in G. The vertex sets of the connected components of $G_c[V \setminus S_i]$ are the same as those of $G[V \setminus S_i]$ $(i = 1, 2)$. It follows that S_1 and S_2 are also non-crossing in G. □

Definition 12.2.2 *Let S_1 and S_2 be two non-crossing separators in G. Consider a connected component D of $G[V \setminus (S_1 \cup S_2)]$. The component D is called between S_1 and S_2, if $S_2 \setminus S_1$ and D are in the same connected component of $G[V \setminus S_1]$ and $S_1 \setminus S_2$ and D are in the same connected component of $G[V \setminus S_2]$.*

We adopt the convention that if $S_2 \subseteq S_1$, then *every* connected component of $G[V \setminus (S_1 \cup S_2)]$ is in the same connected component of $G[V \setminus S_1]$ as $S_2 \setminus S_1$.

Definition 12.2.3 *Let S_1 and S_2 be non-crossing separators in G. The piece $P = P(S_1, S_2)$ is the set of vertices of S_1, S_2 and of all connected components of $G[V \setminus (S_1 \cup S_2)]$ that are between S_1 and S_2.*

For example, notice that if $S_1 = S_2$ then the piece $P(S_1, S_2) = V$. If $S_1 \subset S_2$, the piece consists of S_1 and the vertices of the connected component of $G[V \setminus S_1]$ that contain the vertices of $S_2 \setminus S_1$.

Lemma 12.2.3 *Let S_1 and S_2 be non-crossing separators in G. Let G_C be an efficient triangulation of G such that $S_1, S_2 \in C$. Then the pieces of S_1 and S_2 in G and in G_C are equal.*

Proof. Clearly, the piece in G is a subset of the piece in G_C. Let D be a connected component of $G_C[V \setminus (S_1 \cup S_2)]$ that is in the piece in G_C. Since G_C is efficient, the vertex sets of the connected components of $G_C[V \setminus S_1]$ and $G[V \setminus S_1]$ are the same. Hence D is also contained in the same connected component as $S_2 \setminus S_1$ in $G[V \setminus S_1]$. In the same manner it follows that D and $S_1 \setminus S_2$ are contained in the same connected component of $G[V \setminus S_2]$. It follows that D is contained in the piece in G. □

We have shown that the pieces of S_1 and S_2 in G and in G_C are equal. On the other hand, in general, it is *not* true that the vertex sets of the connected components of $G[V \setminus (S_1 \cup S_2)]$ and $G_C[V \setminus (S_1 \cup S_2)]$ are equal.

Definition 12.2.4 *Let $P = P(S_1, S_2)$ be a piece. The realizer R(P) of P is the graph obtained from G[P] by adding all edges between nonadjacent vertices of S_1 and all edges between non adjacent vertices in S_2.*

Hence in the realizer, both subsets S_i are cliques.

12.3 Pieces $P(S_1, S_2)$ with $S_1 \not\subseteq S_2$ and $S_2 \not\subseteq S_1$

Consider a piece $P = P(S_1, S_2)$ with realizer R(P). Assume there is an efficient triangulation G_C with $S_1, S_2 \in C$ which is an interval graph. In this section we assume that $S_1 \not\subseteq S_2$ and $S_2 \not\subseteq S_1$.

Assume $G_C[P]$ is not a clique, and let x and y be non adjacent vertices in $G_C[P]$. There is a minimal x, y-separator S^* in G_C.

In this section we show that S^* decomposes the piece P into smaller pieces, blocks (which are pieces with one separator contained in another one) and S_1 and S_2. Blocks are treated in the next section.

Lemma 12.3.1 *Using the notation described above:*

1. $S^* \subset P$.

2. S_1 *and* S^* *(and also* S_2 *and* S^* *) are non-crossing in* G.

3. $S^* \neq S_1$ *and* $S^* \neq S_2$.

Proof. By Lemma 12.2.3 the pieces of S_1 and S_2 in G and G_C are equal which enables us to analyze G_C instead G. By assumption $S_2 \setminus S_1 \neq \emptyset$. Consider the connected component C of $G_C[V \setminus S_1]$ that contains $S_2 \setminus S_1$. Then x and y are both contained in $S_1 \cup C \supseteq P$. It follows by Lemma 12.1.1 that S^* is also contained in $S_1 \cup C$. Hence $S^* \setminus (S_1 \cup S_2)$ and $S_2 \setminus S_1$ are in the same connected component. In the same way it follows that $S^* \setminus (S_1 \cup S_2)$ and $S_1 \setminus S_2$ are in the same connected component of $G_C[V \setminus S_2]$ It follows that $S^* \subset P$.

Since S_1, S_2 and S^* are minimal separators in G_C they are pairwise non-crossing in G by Lemma 12.2.2.

Let C be the connected component of $G_C[V \setminus S_1]$ that contains $S_2 \setminus S_1$. Then x and y are both contained in $S_1 \cup C$. It follows that S_1 cannot be a minimal x, y-separator. Hence $S_1 \neq S^*$. $\square$

Lemma 12.3.2 *Assume* $S_1 \subset S^*$ *and* $S_2 \subset S^*$. *There exist connected components* $D_1, \ldots, D_t$ *of* $G_C[V \setminus S^*]$ *which partition* $P \setminus S^*$.

Proof. Let D be a connected component of $G_C[V \setminus S^*]$. We claim that either $D \subset P$ or $D \cap P = \emptyset$. Indeed, notice that D is connected in $G_C[V \setminus (S_1 \cup S_2)]$. $\square$

Lemma 12.3.3 *Let* $S_1 \subset S^*$ *and* $S_2 \setminus S^* \neq \emptyset$. *Then there are connected components* $D_1, \ldots, D_t$ *of* $G_C[V \setminus S^*]$ *such that* $P \setminus S^*$ *can be partitioned into* $\mathcal{P}(S_2, S^*)$ *and* $D_1, \ldots, D_t$.

Proof. Consider the connected components of $G_C[V \setminus S^*]$. One of these, say A, contains $S_2 \setminus S^*$. All other components are either completely contained in P or disjoint from it.

Let $z \in P \cap A$. We show that $z \in \mathcal{P}(S_2, S^*)$. If $z \in S_2 \setminus S^*$ this is clear, hence assume $z \in A \setminus S_2$. Since $z \in P$, it is contained in the connected component of $G_C[V \setminus S_2]$ that contains $S_1 \setminus S_2$. But $S^* \subset P$, hence also $S^* \setminus S_2$ is in this component. Hence z is in the component of $G_C[V \setminus S_2]$ that contains $S^* \setminus S_2$. Since z is also in the component of $G_C[V \setminus S^*]$ that contains $S_2 \setminus S^*$, it follows that $z \in \mathcal{P}(S_2, S^*)$.

Finally we have to show that $\mathcal{P}(S^*, S_2) \subseteq P$. Let $z \in \mathcal{P}(S^*, S_2)$. If $z \in S^* \cup S_2$ then clearly $z \in P$. Hence assume $z \notin S^* \cup S_2$. Then $z \in A$. Since $S_1 \subset S^*$, A is contained in the connected component of $G_C[V \setminus S_1]$ that contains $S_2 \setminus S_1$.

Also, z is in the connected components of $G_C[V \setminus S_2]$ that contains $S^* \setminus S_2$. This component also contains $S_1 \setminus S_2$. Consequently, $z \in \mathcal{P}(S_1, S_2)$ holds.

This proves the lemma. □

The following lemma will be useful. Here we explicitly use the fact that G_C is an interval graph.

Lemma 12.3.4 *Assume $S_1 \not\subseteq S^*$ and $S_2 \not\subseteq S^*$. Then S^* separates $S_1 \setminus S^*$ and $S_2 \setminus S^*$ in G_C.*

Proof. Since G_C is an interval graph, there is an interval ordering of the maximal cliques of G_C, say $C_1, \ldots, C_t$. By Lemma 12.1.3 there are indices i and j such that $S_1 = C_i \cap C_{i+1}$ and $S_2 = C_j \cap C_{j+1}$. Assume $i < j$. Then the piece of S_1 and S_2 is contained in $\bigcup_{k=i+1}^{j} C_k$. The vertices x and y are in this piece. Hence there is an index $i < k < j$ such that $S^* = C_k \cap C_{k+1}$. Consequently, $S_1 \not\subseteq S^*$ and $S_2 \not\subseteq S^*$ implies that S^* separates $S_1 \setminus S^*$ and $S_2 \setminus S^*$. □

Lemma 12.3.5 *Assume $S^* \subset S_1$ and $S^* \subset S_2$. Then $P = S_1 \cup S_2$.*

Proof. For $i = 1, 2$ let D_i be the connected component of $G_C[V \setminus S^*]$ that contains $S_i \setminus S^*$. Notice that the connected component of $G_C[V \setminus S_1]$ that contains $S_2 \setminus S_1$ is just D_2. Hence $P \setminus (S_1 \cup S_2) = D_1 \cap D_2 = \emptyset$. □

Lemma 12.3.6 *Assume $S^* \subset S_1$, $S^* \not\subseteq S_2$ and $S_2 \not\subseteq S^*$. Then $P = S_1 \cup \mathcal{P}(S_2, S^*)$.*

Proof. For $i = 1, 2$ let D_i be the connected component of $G_C[V \setminus S^*]$ that contains $S_i \setminus S^*$. The connected component of $G_C[V \setminus S_1]$ that contains $S_2 \setminus S_1$ is D_2. It follows that $P \setminus S_1 \subseteq D_2$.

Let $z \in P \setminus S_1$. We show that $z \in \mathcal{P}(S^*, S_2)$. By definition z is in the component of $G_C[V \setminus S_2]$ that contains $S_1 \setminus S_2$. This component also contains $S^* \setminus S_2$. Since $z \in D_2$ it follows that $z \in \mathcal{P}(S^*, S_2)$.

It remains to show that $\mathcal{P}(S^*, S_2) \subseteq P$. Let $z \in \mathcal{P}(S^*, S_2) \setminus (S^* \cup S_2)$. Then z is in D_2. Hence z is in the component of $G_C[V \setminus S_1]$ that contains $S_2 \setminus S_1$.

Furthermore, z belongs to the connected component of $G_c[V \setminus S_2]$ that contains $S^* \setminus S_2$. Since $S^* \setminus S_2 \neq \emptyset$, this component is uniquely determined and contains also $S_1 \setminus S_2$, since S_1 is a clique containing S^*. Consequently, $z \in \mathcal{P}(S_1, S_2)$ holds. $\qquad\qquad\qquad\qquad\qquad\qquad\qquad\qquad\qquad\qquad\qquad\qquad\qquad\qquad\qquad\quad\Box$

Lemma 12.3.7 *Assume for* $i = 1, 2$ $S_i \not\subset S^*$ *and* $S^* \not\subset S_i$. *Then there are connected components* $D_1, \ldots, D_t$ *of* $G_c[V \setminus S^*]$ *such that P is partitioned into* $\mathcal{P}(S_1, S^*)$, $\mathcal{P}(S_2, S^*)$ *and* $D_1, \ldots, D_t$.

Proof. Let A and B be the connected components of $G_c[V \setminus S^*]$ which contain $S_1 \setminus S^*$ and $S_2 \setminus S^*$, respectively. Then every other connected component of $G_c[V \setminus S^*]$ is either a subset of P or disjoint from P.

Now let $z \in A \cap P$. We show that $z \in \mathcal{P}(S_1, S^*)$. Since z and $S^* \setminus S_1$ are both in P, and since $S^* \setminus S_1 \neq \emptyset$, it follows that z and $S^* \setminus S_1$ are contained in the same connected component of $G_c[V \setminus S_1]$. Since also $z \in A$ it follows that $z \in \mathcal{P}(S_1, S^*)$.

We now show that $\mathcal{P}(S_1, S^*) \subset P$. Let $z \in \mathcal{P}(S_1, S^*)$. Since $S_1 \setminus S^* \neq \emptyset$ it follows that $z \in S^* \cup A$. If $z \in S^* \cup S_1$ then $z \in P$. Hence we may assume that $z \in A \setminus S_1$.

Now z and $S^* \setminus S_1$ are in the same connected component of $G_c[V \setminus S_1]$ since $z \in \mathcal{P}(S_1, S^*)$. Since $S^* \subset P$ and $S^* \setminus S_1 \neq \emptyset$, this component also contains $S_2 \setminus S_1$. It follows that z and $S_2 \setminus S_1$ are in the same connected component of $G_c[V \setminus S_1]$. This show that $z \in P$ holds. It follows that $A \cap P = \mathcal{P}(S^*, S_1) \setminus S^*$.

$B \cap P = \mathcal{P}(S_2, S^*) \setminus S^*$ can be shown analogously. $\qquad\qquad\qquad\quad\Box$

Notice that in all cases, the partition is such that the constituents are (strictly) smaller than the original piece.

12.4   Pieces $\mathcal{P}(S_1, S_2)$ with $S_1 \subseteq S_2$

In this section let S_1 and S_2 be non-crossing minimal separators in G with $S_1 \subseteq S_2$. Consider the piece $P = \mathcal{P}(S_1, S_2)$ and the realizer $R(P)$. We show how to compute the treewidth of the realizer.

First assume $S_1 \neq S_2$. Then the piece consists of a minimal separator S_1 and the connected component of $G_c[V \setminus S_1]$ that contains $S_2 \setminus S_1$. Notice that in this case, the piece can be partitioned into connected components of $G_c[V \setminus S_2]$.

Now we consider the case $S_1 = S_2$ and denote $S_1 = S_2$ by S. In this case the piece is equal to the total vertex set and the realizer is obtained from G by making a clique of S.

Definition 12.4.1 *A block is a pair* $B = (S, C)$, *where S is a minimal separator of G and C is a connected component of* $G[V \setminus S]$. *The graph* *obtained from* $G[S \cup C]$ *by making a clique of* S *is called the* realizer *of the block and is denoted by* $R(B)$.

Clearly if we can find the treewidth of all realizers of blocks, then this gives us the treewidth of the total graph: assign to each minimal separator a weight which is the maximum treewidth over all realizers incident with this separators. The treewidth of the graph is equal to the minimum weight over all minimal separators.

   Let G_C be an efficient triangulation and let $B = (S, C)$ be a block with realizer R with $S \in \mathcal{C}$. Let x and y be non adjacent vertices in $G_C[S \cup C]$. Let S^* be a minimal x, y-separator in G_C, thus S is also a clique in G_C. Then $S^* \subset S \cup C$ by Lemma 12.1.1.

Lemma 12.4.1 $S^* \neq S$ *and if* $S^* \subset S$ *then* S^* *separates* $S \setminus S^*$ *and* C *in* G_C.

Proof. Assume $S^* \subseteq S$. If x and y are both contained in C then S^* cannot be a minimal x, y-separator, since C is connected in $G_C[V \setminus S*]$. Thus, w.l.o.g. x is contained in $S \setminus S^*$ and y is contained in C. It follows that $S^* \neq S$.

   Now assume some vertex $z \in S \setminus S^*$ has a neighbor in C. Then there is a path from x to y which avoids S^*. $\square$

Lemma 12.4.2 *If* $S \subset S^*$, *then there are connected components* $C_1, \ldots, C_t$ *of* $G_C[V \setminus S^*]$ *which partition* $C \setminus S^*$.

Proof. Obvious. $\square$

Lemma 12.4.3 *Assume* $S \not\subset S^*$ *and* $S^* \not\subset S$. *Then there exist connected components* $C_1, \ldots, C_t$ *of* $G_C[V \setminus S^*]$ *such that* $S \cup C$ *can be partitioned into* $\mathcal{P}(S, S^*)$ *and* $C_1, \ldots C_t$.

Proof. First we show that $\mathcal{P}(S, S^*) \subset S \cup C$. Let $z \in \mathcal{P}(S, S^*)$. We may assume $z \notin S$. Then z and $S^* \setminus S$ are in the same connected component of $G_C[V \setminus S]$. Since $S^* \setminus S \neq \emptyset$, and since $S^* \subset S \cup C$, it follows that $z \in C$. Since $\mathcal{P}(S, S^*)$ cannot both contain x and y (since $S \setminus S^* \neq \emptyset$), it follows that $\mathcal{P}(S, S^*) \neq S \cup C$.

   Since $S \not\subset S^*$ there is exactly one connected components of $G_C[V \setminus S^*]$ that contains $S \setminus S^*$. Moreover, $\mathcal{P}(S, S^*) \setminus S^*$ is contained in that component. Let A be a component of $G_C[C \setminus S^*]$. Assume that A is contained in a connected component B of $G_C[V \setminus S^*]$. If $A \neq B$, then B must contain vertices of $S \setminus S^*$. In that case however, $A \subset \mathcal{P}(S, S^*)$. $\square$

12.5   The algorithm

Using the result of chapter 11 we can find all minimal separators in the graph. For each pair of non-crossing separators, we can compute the piece. We also compute the blocks for every separator. We sort the pieces and the blocks according to increasing number of vertices. Blocks and pieces with the same number of vertices are ordered as follows. Blocks appear in the ordering before pieces with the same number of vertices.

For each block and piece in turn, we compute the treewidth of the realizer, by trying all possible separators that are contained in it, using the results of sections 12.3 and 12.4. If the treewidth of each piece is determined, we look for the piece with vertex set V, with minimum treewidth. This is equal to the treewidth of the graph.

Theorem 12.5.1 *For each constant* d *there exists a polynomial time algorithm that computes the treewidth and pathwidth of cocomparability graphs of dimension at most* d.

Proof. Let R be the number of separators. In chapter 9 it is shown that $R \leq n^d$. Moreover, in chapter 11 it is shown that the set of all separators can be computed in $O(n^5 R)$ time. There are at most R^2 pieces, since these are fully characterized by two minimal separators. For each of these pieces, we can try all minimal separators to split up the piece. For each smaller piece we can look up its treewidth in $O(n^2)$ time. It follows that we can find the treewidth of a piece in $O(Rn^3)$ time. $\square$

Chapter 13

Pathwidth of pathwidth-bounded graphs

The problem of determining the treewidth or pathwidth of a graph, and find-ing tree-decompositions or path-decompositions with small (optimal) width are NP-hard, but if k is a fixed constant the problems are solvable in poly-nomial time. The first algorithms solving these problems for fixed k are based on dynamic programming and use $O(n^{k+2})$ [4] and $O(n^{2k^2+4k+8})$ [57] time, respectively. With help of graph minor theory, Robertson and Sey-mour showed the existence of $O(n^2)$ algorithms. However, these algorithms are non-constructive in two ways: first, only existence of the algorithms is proved and secondly the claimed algorithm gives an output *yes* or *no*, but no tree- or path-decomposition. Another drawback is that the constant factors in the estimated running times of these algorithms make them infeasible. It should be remarked that for $k = 1, 2$ and 3 fast linear time algorithms exist.

In this chapter we show the following result: Given constants A and B, we exhibit a linear time algorithm which, given a tree-decomposition of width at most A, outputs a path-decomposition of width at most B, or correctly states that such a path-decomposition does not exist. Using the approximation algorithm of Reed [139], which finds an approximate tree-decomposition of bounded width in $O(n \log n)$ time, we can then obtain an optimal path-decomposition in $O(n \log n)$ time. Recently Reed's result has been improved by H. Bodlaender [21]. Using the result of the next chapter he shows that for each k there exists a linear time algorithm to decide whether or not the treewidth of a given graph is at most k. We describe this algorithm in chapter 15. Using Bodlaender's result, (finding first a tree-decomposition) the algorithm of this chapter can be used to find for each constant k in linear time if the pathwidth of a given graph is at most k, and if so to determine an optimal path-decomposition for the graph. Recently, the result of this chapter was slightly improved. It was shown in [29] that even for graphs with pathwidth $O(\log n)$, the methods of this chapter can be used to obtain the exact pathwidth. It follows that there is a polynomial algorithm to determine the pathwidth of bounded-treewidth graphs.

Our algorithm does not use non constructive arguments or graph minors but is given directly and needs only constructive combinatorial arguments to prove its correctness. The constant factor, although still growing fast with k is only singly exponential in k.

13.1 Preliminaries

In this section we introduce some extra terminology related to tree-decompositions.

Definition 13.1.1 *A* rooted *tree-decomposition is a tree-decomposition* $D = (S, T)$ *in which* T *is a rooted tree.*

Definition 13.1.2 *Let* $D = (S, T)$ *be a rooted tree-decomposition for a graph* G. *For each node* i *of* T, *let* T_i *be the subtree of* T, *rooted at node* i. *Define:* $V_i = \bigcup_{f \in T_i} X_f$ *and let* $G_i = G[V_i]$ *(so if* r *is the root of* T, $G_r = G$). *We call* G_i *the subgraph of* G *rooted at* i.

We can obtain a rooted tree-decomposition $D_i = (S_i, T_i)$ for G_i from D:

Definition 13.1.3 *Let* $D = (S, T)$ *be a rooted tree-decomposition for a graph* G. *Let* i *be a node of* T. *Let* $D_i = (S_i, T_i)$, *where* T_i *is the subtree of* T *rooted at* i, *and* $S_i = \{X_f \mid f \in T_i\}$. *We call* D_i *the rooted tree-decomposition of* G_i *rooted at node* i.

Lemma 13.1.1 *For each node* i, D_i *is a tree-decomposition for* G_i.

Proof. Clearly, every vertex of G_i is in at least one subset of S_i. To check the second condition for tree-decompositions, let (v, w) be an edge in G_i. There exists a subset X_k in S which contains both end vertices. Assume $X_k \notin S_i$, i.e., k is not a descendant of i in T. But v also is contained in a subset X_p with p a descendant of i. But then, by the third condition of tree-decompositions, v must also be contained in X_i and the same holds for w. Hence the second condition is satisfied. Finally, if T' is a subtree of T, then the restriction of T' to the vertices of T_i is a subtree of T_i, hence a fortiori, also the third condition is satisfied.   □

In order to describe our algorithms more easily, we introduce a special type of rooted tree-decompositions.

Definition 13.1.4 *A rooted tree-decomposition* $D = (S, T)$ *with* $S = \{X_i \mid i \in I\}$ *and* $T = (I, F))$ *is called a* nice *tree-decomposition, if the following conditions are satisfied:*

1. *every node of* T *has at most two children,*

2. *if a node* i *has two children* j *and* k, *then* $X_i = X_j = X_k$

3. *if a node* i *has one child* j, *then either* $|X_i| = |X_j| + 1$ *and* $X_j \subset X_i$ *or* $|X_i| = |X_j| - 1$ *and* $X_i \subset X_j$.

We show that every partial k-tree has a nice tree-decomposition of width k.

Lemma 13.1.2 *Every graph* G *with treewidth* k *has a nice tree-decomposition of width* k. *Furthermore, if* n *is the number of vertices of* G *then there exists a nice tree-decomposition with at most* 4n *nodes.*

Proof. Since G has treewidth k, G has at least $k + 1$ vertices and there is a triangulation of G into a k-tree H. We show that there is a nice tree-decomposition for H of width k. Let n be the number of vertices of H. If $n = k+1$, we can take the trivial tree-decomposition which has one node and the corresponding subset containing all vertices. Assume $n > k+1$. Then H has a simplicial vertex x. Let H' be the k-tree obtained from H by removing x. By induction there is a nice tree-decomposition $(\{X_i | i \in I\}, T = (I, F))$ for H' with at most $4(n - 1)$ nodes. Let C be the set of neighbors of x. Since C is a clique (with k vertices) there exists a subset X_i in the nice tree-decomposition containing C. Take such a node i. We consider three cases.

case 1 First assume i has two children j and k. Then $X_i = X_j = X_k$. We then go down in the tree to either one of the children. We continue until we end in a node p with at most one child.

case 2 If p is a leaf, we add one or two vertices below p in the following manner. If $X_p = C$ we take a new node $a \notin I$ and a subset $X_a = C \cup \{x\}$ and make a a child of p. Otherwise, if $X_p \neq C$, let $z \in X_p \setminus C$. Then we give p a new child a and we create a subset $X_a = X_p \setminus \{z\}$. Notice that since $C \subset X_p$ and $k = |C| \leq |X_p| \leq k + 1$, $X_a = C$. We take another new node b and a subset $X_b = C \cup \{x\}$ and make b a child of a. Hence the number of new vertices introduced when p is a leaf is at most two.

case 3 Finally assume p has one child q. Remove the edge (p, q) from the tree. Take a new node a and corresponding subset $X_a = X_p$. Make a a child of p and q a child of a. Take another new node b, make a new subset $X_b = X_p$ and make b a child of p in the tree (thus p has now two children a and b). We now go down in the tree to b which is a leaf, and continue as described in case 2. Notice that in this case we introduced at most 4 new nodes.

Since the tree-decomposition for H' has at most $4(n-1)$ nodes, the tree-decomposition for G has at most $4n$ nodes.   □

Lemma 13.1.3 *For constant* k, *given a tree-decomposition of a graph* G *of width* k *and* $O(n)$ *nodes, where* n *is the number of vertices of* G, *one can find a nice tree-decomposition of* G *of width* k *and with at most* $4n$ *nodes in* $O(n)$ *time.*

Proof. The construction of the triangulated graph implied by the tree-decomposition clearly takes linear time. Given a triangulated graph, one can find in linear time a perfect elimination scheme (see Lemma 2.1.1, page 8). The lemma now easily follows from the constructive proof of Lemma 13.1.2.

□

Definition 13.1.5 *In a nice tree-decomposition* $(\{X_i | i \in I\}, T = (I, F))$ *every node is of one of four possible types. We name the types as follows.*

'Start' *If a node is a leaf, it is called a* start node.

'Join' *If a node has two children, it is called a* join node.

'Forget' *If a node* i *has one child* j *and if* $|X_i| < |X_j|$, *node* i *is called a* forget node.

'Introduce' *If a node* i *has one child* j *and if* $|X_i| > |X_j|$, *node* i *is called an* introduce node.

Notice that every node in the nice tree-decomposition must have one of the four mentioned labels.

Our algorithm for determining the pathwidth of graphs with small treewidth roughly work as follows.

First we make a (nice) tree-decomposition. Starting with the leaf nodes we work our way up in the tree, at each node computing a table of characteristics of path-decompositions for the subgraph thus far encountered. In each node we calculate the table of partial solutions from the tables at the children of this node.

The main idea is to characterize partial solutions mainly by the vertices that are in the present node.

Notice that for this algorithm it is of crucial importance that we can compute a (nice) tree-decomposition for the graph with small treewidth. Recently Reed obtained the following result [139]:

Lemma 13.1.4 *For each* k *there exists an* $O(n \log n)$ *algorithm that, given a graph* $G = (V, E)$ *as input either finds a tree-decomposition of* G *with treewidth* $\leq 4k$ *or determines that the treewidth of* G *is larger than* k.

This improves an earlier result of Lagergren, who gave an $O(n \log^2 n)$ algorithm to find a tree-decomposition of width at most $6k + 5$ if the treewidth of G is at most k [115].

The first step of our algorithm is to change this approximate tree-decomposition found by the algorithm of Reed into a nice tree-decomposition, which can be done in linear time. In the next section we assume that we have a nice tree-decomposition and we show how to use this in order to find the pathwidth of the graph.

13.2 A decision algorithm for pathwidth

In this section we assume that we have a nice tree-decomposition $D = (S, T)$ with $S = \{X_i | i \in I\}$ and $T = (I, F)$, of width k of the input graph $G = (V, E)$. We give an algorithm to decide whether the pathwidth of G is at most ℓ. k and ℓ are considered to be constants for the algorithm. We use a somewhat more explicit notation for a path-decomposition of a graph, i.e., we use the notation $P = (Z_1, Z_2, \ldots, Z_m)$ (Definition 2.2.2, page 18). Recall Definition 13.1.3 for the tree-decomposition D_i rooted at node $i \in I$. We similarly define a partial path-decomposition rooted at node i.

Definition 13.2.1 *A partial path-decomposition rooted at node* $i \in I$ *is a path-decomposition for* G_i, *the subgraph of* G *rooted at* i.

Our algorithm roughly works as follows. For each node i we define a *full set of characteristics* for partial path-decompositions rooted at i of width at most ℓ. By this we mean:

> If there exists a partial path-decomposition of width at most ℓ, there also exists a partial path-decomposition of which the characteristic is in the full set of characteristics.

Assuming that we have a full set of characteristics for each of the children of a node i, we show how to compute the full set of characteristics for node i. We also show that a full set of characteristics is of *bounded* size, and that the full set of characteristics at node i can be computed in $O(1)$ time from the sets at the children of i. If we have computed the full set of characteristics for the root r of T, we can decide whether the pathwidth of G is at most ℓ: this is the case if and only if the full set of characteristics is not empty. In computing the full set of characteristics for node i we consider four different

cases, namely corresponding to whether i is a **start, join, forget** or **introduce**
node. Notice that, if k and ℓ are constants and the tree-decomposition D of
width k and with a linear number of nodes is given, the algorithm to decide
whether the pathwidth of G is at most ℓ takes only linear time.

13.3   The interval model

The characteristic of a partial path-decomposition is a pair of which the first
element is called the *interval model* of the path-decomposition.

Definition 13.3.1 *Let* $Y = (Y_1, \ldots, Y_r)$ *be a partial path-decomposition
rooted at node i. The* restriction of Y *is the sub-decomposition* Y^* *of* Y
for the subgraph induced by X_i, *i.e.* $Y^* = (Y_1 \cap X_i, \ldots, Y_r \cap X_i)$.

In the restriction Y^* there can be many consecutive elements which are the
same. If we remove these duplicates, we obtain the interval model for Y
which is, of course, still a path-decomposition for the subgraph induced by
X_i.

Definition 13.3.2 *Let* $Y^* = (Z_1, \ldots, Z_r)$ *be the restriction of a path-
decomposition* Y *rooted at i. Let* $1 = t_1 < \ldots < t_{q+1} = r + 1$ *be defined
by:*

$$\forall_{1 \leq i \leq q} \forall_{t_i \leq s < t_{i+1}} [Z_s = Z_{t_i}] \wedge \forall_{1 \leq i < q} [Z_{t_{i+1}} \neq Z_{t_i}]$$

The interval model *for* Y *at node i is the sequence* $(Z_{t_i})_{1 \leq i \leq q}$.

Notice that *not every* path-decomposition for a subgraph induced by X_p
without repeating subsets is an interval model, since an interval model is
defined by means of a partial path-decomposition rooted at p. We call a
path-decomposition for the subgraph X_p without adjacent subsets that are
the same, minimal:

Definition 13.3.3 *A path-decomposition Z for a graph G is called* minimal *if no two consecutive subsets in Z are the same.*

The next lemma shows that there are only $\Theta(1)$ different interval models at
each node i.

Lemma 13.3.1 *For each node i the number of different interval models
at i is bounded by* $(2k + 3)^{2k+3}$. *The number of subsets in any interval
model is at most* $2k + 3$. *These bounds hold for the minimal path-
decompositions for the subgraph induced by* X_i *as well.*

Proof. An interval model at node i is a path-decomposition $Z = (Z_1, \ldots, Z_r)$
for $G[X_i]$ which is minimal. We show the bounds hold for the minimal path-
decompositions. Let $L(s)$ be the maximal number of subsets in a minimal

path-decomposition of a graph with s vertices. We claim that $L(s) \leq 2s+1$. Clearly, $L(1) = 3$. Now let $s > 1$, and let $Z = (Z_1, \ldots, Z_r)$ be a minimal path-decomposition for a graph $H = (V, E)$ with s vertices. Take any vertex x and let Z_a and Z_b be the first and the last subset of Z containing x. Now remove x from the graph and let $H' = H[V \setminus \{x\}]$. We can obtain a path-decomposition Z' for H' by removing vertex x from all subsets of Z. Notice that Z' can have at most two pairs of duplicate subsets, namely Z'_a can be the same as Z'_{a-1} and Z'_b can be the same as Z'_{b+1}. It follows that the number of subsets of Z is at most two more than the maximal number of subsets in a path-decomposition of a graph with $s - 1$ vertices. Hence $L(s) \leq L(s - 1) + 2$. This proves our claim. Since $|X_i| \leq k + 1$, the number of subsets in a minimal path-decomposition of $G[X_i]$ is at most $2k + 3$.

We can find an upperbound for the number of interval models as follows. Notice that an interval model can be characterized by indicating for each vertex the first and last subset where it is contained in. Thus we find an upper bound of ι^{2k+2} for the number of interval models with i subsets. Hence we find:

$$\text{number of interval models} \leq \sum_{i=1}^{2k+3} \iota^{2k+2} \leq (2k + 3)^{2k+3}$$

This proves the lemma. □

13.4 Typical sequences

Consider a partial path-decomposition $Y = (Y_1, \ldots, Y_r)$ rooted at i of width at most ℓ. Let Z be the restriction of Y, i.e.,

$$\forall_{1 \leq j \leq r} [Z_j = Y_j \cap X_i]$$

In the previous section we defined the interval model of Y as a subsequence $(Z_{t_i})_{1 \leq i \leq q}$ of Z. For each *interval* $[t_i, t_{i+1})$ consider the integer sequence $(|Y_{t_i}|, |Y_{t_i+1}|, \ldots, |Y_{t_{i+1}-1}|)$. In this subsection we show how to characterize these integer sequences by *typical sequences*. We show that these typical sequences (which are also integer sequences) are bounded in length by $4\ell + 7$ and that the maximum element of these typical sequences is bounded by $\ell+1$. In the next section we define the characteristic of the path-decomposition Y as the pair $((Z_{t_i})_{1 \leq i \leq q}, (\delta^{(i)})_{1 \leq i \leq q})$, where each $\delta^{(i)}$ is the typical sequence for $(|Y_{t_i}|, |Y_{t_i+1}|, \ldots, |Y_{t_{i+1}-1}|)$.

In this section we give some results on nonnegative integer sequences, of which the maximum element is bounded by some constant L (we use $L = l+1$). When we say 'integer sequence' we shall always mean 'nonnegative integer sequence' of length at least one. We start with some notations.

- For any integer sequence $a(1\ldots n)$, let $l(a) = n$ be the length and $max(a)$ be the maximum value: $max(a) = max_{1 \leq i \leq n} a_i$.

- For two sequences a and b *of the same length* we define the sum $c = a + b$ as the sequence c with

$$\forall_{1 \leq i \leq l(a)} \left[c_i = a_i + b_i \right]$$

- For two sequences a and b *of the same length* we write $a \leq b$ if $\forall_i \ a_i \leq b_i$.

- For a *constant* A we write $a + A$ for the sequence with $\forall_i \ (a + A)_i = a_i + A$.

Definition 13.4.1 *For an integer sequence $a(1\ldots n)$ we define the typical sequence $\tau(a)$ as the sequence obtained after iterating the following operations.*

- *Remove consecutive repetitions of the same element, i.e.   if $a_i = a_{i+1}$ then the sequence $a = (a_1, \ldots, a_n)$ is replaced by $(a_1, \ldots, a_i, a_{i+2}, \ldots, a_n)$.*

- *If the sequence contains two elements a_i and a_j such that $j - i > 2$ and $\forall_{i < k < j} \left[a_k \geq max(a_i, a_j) \right]$, then replace the subsequence $a(i + 1 \ldots j - 1)$ by one element equal to its maximum.*

The limit sequence is $\tau(a)$. We refer to the second operation as the typical operation.

If a typical operation is applied to a subsequence $a(i\ldots j)$ of $a = (a_1, \ldots, a_n)$, then a new sequence

$$a' = (a_1, \ldots, a_i, a_k, a_j, \ldots, a_n)$$

is constructed where a_k is the maximum element of the subsequence $a(i\ldots j)$. Notice that a' is at least one element shorter than a. By definition, $\tau(a)$ contains no repetitions and the typical operation is not applicable any further.

Lemma 13.4.1 *For every a, the typical sequence $\tau(a)$ is uniquely defined.*

Proof. Consider a sequence $a(1\ldots n)$. An interval (i, j) is *maximal* if the typical operation is applicable to $a(i\ldots j)$, and the interval (i, j) is not contained in a larger interval to which the typical operation is applicable. We claim that the maximal intervals are disjoint (share at most an endpoint).

Assume two maximal intervals (i_1, j_1) and (i_2, j_2) overlap, i.e., $i_1 < i_2 < j_1 < j_2$. Then notice that $a(i_2) \geq a(i_1)$ and $a(j_1) \geq a(j_2)$. Hence:

$$\forall_{i_2 < s < j_2} \left[a(s) \geq a(i_2) \geq a(i_1) \wedge a(s) \geq a(j_2) \right]$$

$$\forall_{i_1 < s < j_1} \left[a(s) \geq a(j_1) \geq a(j_2) \wedge a(s) \geq a(i_1) \right]$$

This shows that the typical operation is also applicable to (i_1, j_2), which is a contradiction. It follows that the order in which the typical operations are applied does not influence the resulting typical sequence. It is easy to see that the removal of repetitions may be applied at any moment. $\square$

Lemma 13.4.2 *Let $a(1 \ldots n)$ be a sequence of nonnegative integers with $\max(a) = L$. Then $\max(\tau(a)) = L$ and $l(\tau(a)) \leq 4L + 3$. The number of different typical sequences of which the maximum element is bounded by L is at most $(L + 1)^{4L+4}$.*

Proof. The first part of the claim is trivial. We prove that the maximal length of any sequence for which the typical operation is not applicable is at most $4L + 3$ (with L the maximum value of the sequence).

Consider sequences that start with a 0 and that do not contain any other 0. Let $\tilde{T}(L)$ be the maximum length of such a sequence. Notice that $\tilde{T}(0) = 1$. The next minimum element (larger than zero) can occur only at the second or third position in the sequence since otherwise the typical operation is applicable. We find the following recurrence relation:

$$\tilde{T}(L) \leq \tilde{T}(L - 1) + 2$$

Hence $\tilde{T}(L) \leq 2L + 1$.

Now let $T(L)$ be the maximum length of a sequence with maximum value L and for which the typical operation is not applicable. Such a sequence can have at most three zeros of which the outermost two can be at distance at most two. Hence:

$$T(L) \leq 2\tilde{T}(L) + 1 \leq 4L + 3$$

The number of different typical sequences with length i is clearly bounded by $(L + 1)^i$. Hence the total number of different typical sequence is at most

$$\text{number of typical sequences} \leq \sum_{i=0}^{4L+3} (L + 1)^i \leq (L + 1)^{4L+4}$$

$\square$

Remark. The following sequence shows that the bound on the length of a typical sequence in Lemma 13.4.2 is optimal:

$$L, L, L - 1, L, L - 2, L, \ldots, 1, L, 0, L, 0, L, 1, L, 2, \ldots, L - 2, L, L - 1, L, L$$

Definition 13.4.2 *Let* $a(1..n)$ *be a sequence. We define* $E(a)$ *as the set of extensions of* a:

$$E(a) = \{a^* \mid \exists_{1=t_1 < t_2 < ... < t_{n+1}} \forall_{1 \le i \le n} \forall_{t_i \le k < t_{i+1}} [\, a^*(k) = a(i)\,]\}$$

Hence each element of $E(a)$ is of the form $(a_1, a_1, \ldots, a_2, a_2, \ldots, a_n, \ldots, a_n)$, where each a_i of the original sequence a appears at least once in the extension. For any interval $[\alpha, \beta]$ with $t_i \le \alpha \le \beta < t_{i+1}$ we say that $a(i)$ is *repeated* in this interval (in a^*).

Lemma 13.4.3 *If* $a^* \in E(a)$ *then* $\tau(a^*) = \tau(a)$.

Proof. In computing $\tau(a^*)$ we may start by removing all repetitions.   □

Definition 13.4.3 *For two integer sequences* a *and* b *we write* $a \prec b$ *if there are* $a^* \in E(a)$ *and* $b^* \in E(b)$ *of the same length such that* $a^* \le b^*$. *If both* $a \prec b$ *and* $b \prec a$ *hold we write* $a \equiv b$.

Lemma 13.4.4 *The relation* $\prec$ *is transitive.*

Proof. Let $a \prec b$ and $b \prec c$. First notice that:

$$b \prec c \wedge b^* \in E(b) \Rightarrow b^* \prec c$$

We show there exist extensions $a^* \in E(a)$ and $b^* \in E(b)$ such that $a^* \le b^*$. By the remark above, $b^* \prec c$ and hence there are extensions $b^{**} \in E(b^*)$ and $c^* \in E(c)$ such that $b^{**} \le c^*$. We make an extension $a^{**} \in E(a^*)$ as follows. If an element of b^* is repeated in b^{**} then we let the corresponding element of a^* repeat in a^{**}. Clearly,

$$a^{**} \le b^{**} \le c^*$$

and hence $a \prec c$.   □

Corollary 13.4.1 *The relation* $\equiv$ *is an equivalence relation.*

Lemma 13.4.5 *If a sequence* a' *is obtained from a sequence* a *by a typical operation then* $a' \equiv a$. *Moreover, there exist extensions* a'^* *and* a'^{**} *both of* a' *such that* $a'^* \le a \le a'^{**}$.

Proof. We first show that $a \prec a'$. Let a' be the sequence obtained by applying the typical operation to the subsequence $a(i+1 \ldots j-1)$. So the subsequence $a(i+1 \ldots j-1)$ is replaced by its maximum element. We can make an extension of a', in which this maximum element is repeated $j-i-1$ times. This shows $a \prec a'$. To see that $a' \prec a$ also holds, consider again the typical operation applied to a subsequence $a(i+1\ldots j-1)$. Let a_k be the maximum element of this subsequence. Now, in a', repeat the element a_i $k-i+1$ times, and element a_j $j-k$ times. This proves $a' \prec a$. Notice that in both cases we only used an extension of a', hence the 'moreover part' of the lemma follows.   □

Lemma 13.4.6 *For any integer sequence* a: $\tau(a) \equiv a$. *Moreover, there exist extensions* a' *and* a'' *of* $\tau(a)$ *and both of the same length as* a *such that* $a'_i \leq a_i$ *and* $a''_i \geq a_i$ *for all* i.

Proof. First assume that a contains no two consecutive elements that are the same. Recall that, in this case, $\tau(a)$ is obtained from a by a series of typical operations. We prove the following: if a' is obtained from a by a series of typical operations then there exists an extension a'^* of a' such that $a'^* \leq a$. We prove this by induction on the number of typical operations applied. If this number is zero the statement follows immediately. Otherwise, consider the last typical operation applied. Let b be the sequence obtained from a before this last typical operation is applied. So a' is obtained from b by one typical operation. By Lemma 13.4.5 there is an extension c of a' such that $c \leq b$. By induction there is an extension d of b such that $d \leq a$. It immediately follows that there also exists an extension a'^* of a' satisfying $a'^* \leq a$: if an element of b is repeated r times in d, repeat the corresponding element in c also r times.

In the same manner one can show that there is an extension a'^{**} of a' such that $a \leq a'^{**}$.

Finally, for any extension of a we can extend a' and a'' in the same manner which proves the lemma.   □

From Lemma 13.4.6 and Lemma 13.4.4 it follows that:

Corollary 13.4.2 *If* a *and* b *are two sequences then* $a \prec b$ *if and only if* $\tau(a) \prec \tau(b)$.

Definition 13.4.4 *Let* $a(1..n)$ *and* $b(1..m)$ *be two integer sequences. The* ringsum $a \oplus b$ *is defined as:*

$$a \oplus b = \{a^* + b^* \mid a^* \in E(a) \text{ and } b^* \in E(b) \text{ and } l(a^*) = l(b^*)\}$$

Lemma 13.4.7 *Let* $c \in a \oplus b$, *and let* $a^* \in E(a)$ *and* $b^* \in E(b)$. *Then there exists a sequence* $c^* \in E(c)$ *such that* $c^* \in a^* \oplus b^*$.

Proof. Let $c = a' + b'$ for some $a' \in E(a)$ and $b' \in E(b)$. Let a_i be repeated p'_i times in a' and p^*_i times in a^*. Let b_i be repeated q'_i times in b' and q^*_i times in b^*. Let $\lambda \geq 1$ be an integer. Make new extensions $a^\circ \in E(a)$ and $b^\circ \in E(b)$, by repeating a_i $\lambda p'_i$ times in a° and b_j $\lambda q'_j$ times in b°. Then a° and b° have the same length. If $c^* = a^\circ + b^\circ$, then $c^* \in E(c)$. If we take λ such that $\lambda p'_i \geq p^*_i$ for all i and $\lambda q'_j \geq q^*_j$ for all j then also $a^\circ \in E(a^*)$ and $b^\circ \in E(b^*)$. $\square$

Next we show that if two sequences can be 'improved', then also the sum can be improved.

Lemma 13.4.8 *Let a and b be two integer sequences of the same length and let $y = a + b$. Let $a_0 \prec a$ and $b_0 \prec b$. Then there is a sequence $y_0 \in a_0 \oplus b_0$ such that $y_0 \prec y$.*

Proof. There are extensions a^*_0 of a_0 and a^* of a such that $a^*_0 \leq a^*$ and extensions b^*_0 of b_0 and b^* of b such that $b^*_0 \leq b^*$. Assume an element a_i is repeated p_i times in a^* and b_i is repeated q_i times in b^*. Now change the extensions a^* and a^*_0 into a^{**} and a^{**}_0 by repeating a_i $p_i q_i$ times in a^{**} and repeating each corresponding element in a^*_0 q_i times. We then have $a^{**}_0 \leq a^{**}$. In a similar way we obtain new extensions b^{**}_0 and b^{**}. Make an extension y^{**} of y by repeating each element y_i $p_i q_i$ times. Define $y_0 = a^{**}_0 + b^{**}_0$. We now have $y_0 = a^{**}_0 + b^{**}_0 \leq a^{**} + b^{**} = y^{**}$ and $y_0 \in a_0 \oplus b_0$. $\square$

Lemma 13.4.9 *Let a and b be two integer sequences and let $c \in a \oplus b$. Then there exists an element $c' \in \tau(a) \oplus \tau(b)$ such that $c' \prec c$.*

Proof. This is an immediate consequence of Lemma 13.4.8. By Definition 13.4.4 $c = a^* + b^*$ for some extensions a^* of a and b^* of b. By Lemma 13.4.3 $\tau(a^*) = \tau(a)$ and $\tau(b^*) = \tau(b)$. By Lemma 13.4.6 $\tau(a^*) \prec a^*$ and $\tau(b^*) \prec b^*$. Hence, by Lemma 13.4.8, there is a $c^* \in \tau(a) \oplus \tau(b)$ such that $c^* \prec c$. $\square$

Definition 13.4.5 *Let $a(1 \ldots n)$ and $b(1 \ldots m)$ be two integer sequences. The* concatenation *of a and b, is defined as the sequence:*

$$a +\!\!+ b = (a_1, \ldots, a_n, b_1, \ldots, b_m)$$

Lemma 13.4.10 *For any two sequences a and b we have: $\tau(a +\!\!+ b) = \tau(\tau(a) +\!\!+ \tau(b))$.*

Proof. By Lemma 13.4.1 we can apply typical operations and removal of duplicates in any order to obtain $\tau(a +\!\!+ b)$. Start by applying the typical operations of a to the sequence $a +\!\!+ b$ and remove adjacent duplicates from this sublist. The result is the sequence $\tau(a) +\!\!+ b$. Next apply all typical operations and removal of duplicates to the sublist b. The result is $\tau(a) +\!\!+ \tau(b)$. This proves the lemma.   □

Lemma 13.4.11 *If $a^* \in E(a)$ and $b^* \in E(b)$ then $a^* +\!\!+ b^* \in E(a +\!\!+ b)$.*

Proof. Obvious.   □

Lemma 13.4.12 *If $a' \prec a$ and $b' \prec b$, then $a' +\!\!+ b' \prec a +\!\!+ b$.*

Proof. There are extensions $a'^* \in E(a')$, $a^* \in E(a)$, $b'^* \in E(b')$ and $b^* \in E(b)$ such that $a'^* \leq a^*$ and $b'^* \leq b^*$. Then clearly also: $(a'^* +\!\!+ b'^*) \leq (a^* +\!\!+ b^*)$. By Lemma 13.4.11: $a^* +\!\!+ b^* \in E(a +\!\!+ b)$ and $a'^* +\!\!+ b'^* \in E(a' +\!\!+ b')$. This proves the lemma.   □

Definition 13.4.6 *Let $a(1 \ldots n)$ be an integer sequence $(n > 0)$. A split of a is a pair (δ_1, δ_2) of integer sequences of one of two types.*

1. *The first type split is such that there exists an index $1 \leq f \leq n$ with: $\delta_1 = (a_1, \ldots, a_f)$ and $\delta_2 = (a_f, \ldots, a_n)$;*

2. *The second type split is such that there is an index $1 \leq f \leq n$ with: $\delta_1 = (a_1, \ldots, a_f)$ and $\delta_2 = (a_{f+1}, \ldots, a_n)$.*

Notice that a_f occurs in both elements of the split of the first type. For an integer sequence of length one there can only be a split of the first type, since we assumed that integer sequences always have length at least one.

Lemma 13.4.13 *Let a be a nonempty sequence such that $a \in E(\tau(a))$. Let (δ_1, δ_2) be a split of $\tau(a)$ of any type. Let (a_1, a_2) be a split of a of the same type such that $a_1 \in E(\delta_1)$ and $a_2 \in E(\delta_2)$ (this split exists). Then $\tau(a_1) = \delta_1$ and $\tau(a_2) = \delta_2$.*

Proof. Write $\tau(a) = (\alpha_1, \ldots, \alpha_s)$ and let (δ_1, δ_2) be a split of the first type with $\delta_1 = (\alpha_1, \ldots, \alpha_f)$ and $\delta_2 = (\alpha_f, \ldots, \alpha_s)$. Make a split of a of the first type such that

$$a_1 = (\alpha_1, \ldots, \alpha_1, \ldots, \alpha_f, \ldots, \alpha_f) \land a_2 = (\alpha_f, \ldots, \alpha_f, \ldots, \alpha_s, \ldots, \alpha_s)$$

(with α_f appearing at least once in each a_i). This split clearly is possible since $a \in E(\tau(a))$. Since $a \in E(\tau(a))$, $\tau(a)$ is obtained from a by removing repetitions of elements in a. Clearly, δ_i contains no repetitions, and no typical operation is applicable to it. If the split (δ_1, δ_2) is of the second type, the proof is similar. Hence the lemma. $\square$

We extend the results on integer sequences to *lists of integer sequences*. We use the notation $[a]$ to represent a list $(a^{(1)}, a^{(2)}, \ldots, a^{(n)})$ where each $a^{(i)}$ represents an integer sequence. For short, we call a list of integer sequences also a list. We start with some notations.

1. The length of a list is the number of integer sequences in the list.

2. For a list $[a] = (a^{(1)}, \ldots, a^{(n)})$ we define $\max([a]) = \max_{1 \le i \le n} \max(a^{(i)})$.

3. For two lists $[a] = (a^{(1)}, \ldots, a^{(n)})$ and $[b] = (b^{(1)}, \ldots, b^{(n)})$ of the same length and such that $l(a^{(i)}) = l(b^{(i)})$ for all i, we say that $[a]$ and $[b]$ have the same length *in the strong sense*.

4. For two lists $[a]$ and $[b]$ with the same length in the strong sense we write $[a] \le [b]$ if $a^{(i)} \le b^{(i)}$ for each i.

5. For two such lists with the same length in the strong sense we use the notation $[a] + [b]$ for the list $(a^{(1)} + b^{(1)}, \ldots, a^{(n)} + b^{(n)})$.

6. Let $[a] = (a^{(1)}, \ldots, a^{(n)})$ be a list. The typical list $\tau[a]$ of $[a]$ is the list

$$\tau[a] = (\tau(a^{(1)}), \ldots, \tau(a^{(n)}))$$

7. Let $[a] = (a^{(1)}, \ldots, a^{(n)})$ be a list. The set of *extensions* of $[a]$ is defined as:

$$E[a] = \{[b] = (b^{(1)}, \ldots, b^{(n)}) \mid \forall_i \, b^{(i)} \in E(a^{(i)})\}$$

8. The *ringsum* of two lists $[a] = (a^{(1)}, \ldots, a^{(n)})$ and $[b] = (b^{(1)}, \ldots, b^{(n)})$ *of the same length* is defined as

$$[a] \oplus [b] = \{(c^{(1)}, \ldots, c^{(n)}) \mid \forall_i \, c^{(i)} \in a^{(i)} \oplus b^{(i)}\}$$

9. For two lists $[a]$ and $[b]$ *of the same length* we write $[a] \prec [b]$ if there exist extensions $[a^*] \in E[a]$ and $[b^*] \in E[b]$ such that $[a^*] \le [b^*]$. If both $[a] \prec [b]$ and $[b] \prec [a]$ we write $[a] \equiv [b]$.

Most results on integer sequences trivially extend to lists of integer sequences. We summarize them in the following lemma.

Lemma 13.4.14

1. *The relation $\prec$ is transitive for lists and $\equiv$ is an equivalence relation for lists (Lemma 13.4.4 and Corollary 13.4.1).*

2. *If $[b] \in E[a]$ then $\tau[b] = \tau[a]$ (Lemma 13.4.3).*

3. *For two lists $[a]$ and $[b]$ of the same length: $[a] \prec [b] \Leftrightarrow \tau[a] \prec \tau[b]$ (Corollary 13.4.2).*

4. *For any list $[a]$: $\tau[a] \equiv [a]$. Moreover there are extensions $[b'] \in E(\tau[a])$ and $[b''] \in E(\tau[a])$ such that $[b'] \leq [a] \leq [b'']$ (Lemma 13.4.6).*

5. *Let $[a]$ and $[b]$ be two lists of the same length and let $[c] \in [a] \oplus [b]$. Let $[a^*] \in E[a]$ and $[b^*] \in E[b]$. Then there exists a list $[c^*] \in E[c]$ such that $[c^*] \in [a^*] \oplus [b^*]$ (Lemma 13.4.7).*

6. *Let $[a]$ and $[b]$ be two list with the same length in the strong sense and let $[y] = [a] + [b]$. Let $[a_0] \prec [a]$ and $[b_0] \prec [b]$. Then there exists a list $[y_0] \in [a_0] \oplus [b_0]$ such that $[y_0] \prec [y]$ (Lemma 13.4.8).*

7. *Let $[a]$ and b be two lists of the same length and let $[c] \in [a] \oplus [b]$. There exists a list $[c'] \in \tau[a] \oplus \tau[b]$ such that $[c'] \prec [c]$ (Lemma 13.4.9).*

13.5 Characteristic path-decompositions

Let $D = (S, T)$ be a nice tree-decomposition with $S = \{X_i | i \in I\}$ and $T = (I, F)$ for the graph $G = (V, E)$ of width k. Consider partial path-decompositions rooted at some node $i \in I$. In this section we define a full set of characteristics of path-decompositions rooted at some node i. We start with defining the characteristic of a path-decomposition.

Definition 13.5.1 *Let Y be a partial path-decomposition rooted at a node i. Let $Z = (Z_{t_j})_{1 \leq j \leq q}$ be the interval model for Y. The list representation for Y is the pair $(Z, [Y])$, where Z is the interval model and $[Y] = (Y^{(1)}, Y^{(2)}, \ldots, Y^{(q)})$ is the sequence with $Y^{(m)} = (Y_{t_m}, Y_{t_m+1}, \ldots, Y_{t_{m+1}-1})$ for each $1 \leq m \leq q$.*

Definition 13.5.2 *For a partial path-decomposition $Y = (Y_1, Y_2, \ldots, Y_m)$, the set of extensions of Y, $E(Y)$, is the set of path-decompositions*

$$Z = (Y_1, Y_1, \ldots, Y_1, Y_2, Y_2, \ldots, Y_2, \ldots, Y_m, \ldots, Y_m)$$

where each subset Y_i is repeated at least once.

Definition 13.5.3 *Let* Y *be a partial path-decomposition with list representation* $(Z, [Y])$, *with interval model* $Z = (Z_{t_j})_{1 \leq j \leq q}$. *Let* $[y] = (y^{(1)}, y^{(2)}, \ldots, y^{(q)})$ *be the list of integer sequences with* $y^{(m)} = (|Y_{t_m}|, |Y_{t_m+1}|, \ldots, |Y_{t_{m+1}-1}|)$ *for each interval* $1 \leq m \leq q$. *We call* $[y]$ *the list of* Y *and* $\tau[y]$ *the typical list of* Y.

Definition 13.5.4 *Let* Y *be a partial path-decomposition with list representation* $(Z, [Y])$ *and let* $[y]$ *be the list of* Y. *The* characteristic *of* Y *is the pair*

$$C(Y) = (Z, \tau[y])$$

Lemma 13.5.1 *The number of different characteristics is bounded by*

$$(2k + 3)^{2k+3}(\ell + 2)^{(4\ell+8)(4k+5)}$$

Proof. By Lemma 13.3.1 and by Lemma 13.4.2. For each subset in the interval model we choose a typical sequence. Since the number of subsets is bounded by $2k + 3$, the number of typical lists with each interval model is at most $(\ell + 2)^{(4\ell+8)(2k+3)}$. $\qquad \Box$

Definition 13.5.5 *For two partial path-decompositions* Y *and* Z *rooted at the same node* i, *which have the same interval model, we write* $Y \prec Z$ *if the corresponding lists satisfy* $[y] \prec [z]$. *If* $Y \prec Z$ *and* $Z \prec Y$, *we write* $Y \equiv Z$.

Definition 13.5.6 *A* set of characteristics *$FS(i)$, of partial path-decompositions rooted at some node* i, *of width at most* ℓ *is called a* full set of characteristics *if for each partial path-decomposition* Y *rooted at* i *of width at most* ℓ *there is a path-decomposition* $Y' \prec Y$ *such that the characteristic of* Y' *is in* $FS(i)$.

Lemma 13.5.2 *If some full set of characteristics at a node* i *is nonempty, then every full set of characteristics at this node is nonempty. A full set of characteristics is nonempty if and only if the pathwidth of* G_i *is at most* ℓ.

Proof. By Definition 13.5.6. $\qquad \Box$

Notice that the pathwidth of G is at most ℓ if and only if any full set of characteristics at the root of the tree-decomposition is non-empty. In the next four subsections we show how to compute a full set of characteristics at a node p in $O(1)$ time, when a full set of characteristics of all the children of p is given.

13.6 A full set for a start node

In this subsection we consider the case in which p is a **start** node of the nice tree-decomposition $D = (S, T)$. Clearly, in this case G_p is the subgraph induced by the subset X_p.

Lemma 13.6.1 *If Y is partial path-decomposition for a* **start** *node p with list representation $(Z, [Y])$, then for each interval $[t_m, t_{m+1})$ the corresponding list of subsets is of the form $Y^{(m)} = (Z_{t_m}, Z_{t_m}, \ldots, Z_{t_m})$ and has length at least one.*

Proof. The restriction (Definition 13.3.1) of the path-decomposition Y rooted at p, is the path-decomposition Y itself, since p is a **start** node. The lemma follows by Definition 13.3.2. □

Lemma 13.6.2 *If Y is a partial path-decomposition for a* **start** *node p, with list representation $(Z, [Y])$, then the partial path-decomposition $Y' = (Z_{t_1}, Z_{t_2}, \ldots, Z_{t_q})$ satisfies $Y' \prec Y$.*

Proof. By Lemma 13.6.1 $Y \in E(Y')$. □

We can compute a full set of characteristics $FS(p)$ for the **start** node p as follows. Make a list of all interval models of p (i.e. all minimal path-decompositions of G_p). By Lemma 13.3.1 the number of these is bounded by a constant and the different interval models can be computed in constant time. For each interval model $Z = (Z_{t_i})_{1 \le i \le q}$ let $[z] = ((|Z_{t_1}|), (|Z_{t_2}|), \ldots, (|Z_{t_q}|))$ be the list of the partial path-decomposition $(Z_{t_1}, \ldots, Z_{t_q})$. Notice that $\tau[z] = z$ (since each integer sequence has only one element). The full set of p consists of all pairs $(Z, \tau[z])$ such that each interval Z_{t_i} contains at most $\ell + 1$ elements (by Lemma 13.6.2 this *is* a full set).

13.7 A full set for a join node

Let p be a **join** node with children q and r. By definition $X_p = X_q = X_r$, since the tree-decomposition is nice. We now have $V_q \cap V_r = X_p$ (V_i is the vertex set of G_i) and G_p is obtained from G_q and G_r by identifying the vertices of G_q and G_r that are in X_p.

We first show how to compute a full set of characteristics of p when full sets of characteristics for q and r are given. Take $(Z, \tau[a]) \in FS(q)$ and $(Z, \tau[b]) \in FS(r)$ with the same interval model $Z = (Z_{t_i})_{1 \le i \le w}$ (thus $\tau[a]$ and $\tau[b]$ have the same length). Let $\tau[a] = (\tau(a^{(1)}), \ldots, \tau(a^{(w)}))$.

First compute the list $[a^*] = (\tau(a^{(1)}) - |Z_{t_1}|, \ldots, \tau(a^{(w)}) - |Z_{t_w}|)$ (we do this in order to prevent counting the elements of Z_{t_i} twice, as they are counted both in $\tau[a]$ and in $\tau[b]$). Compute the set

$$FS(p) = \{(Z, \tau[c]) | (Z, \tau[a]) \in FS(q) \wedge (Z, \tau[b]) \in FS(r) \wedge$$
$$[c] \in [a^*] \oplus \tau[b] \wedge \max([c]) \le \ell + 1\}$$

We show that $FS(p)$ is a full set of characteristics for p. We first show that an element of $FS(p)$ is indeed a characteristic of a partial path-decomposition at node p. The following lemmas will be useful.

Lemma 13.7.1 *Let* Y *be a partial path-decomposition. Let* $[y]$ *be the list of* Y. *If* $[y^*] \in E[y]$, *then there exists a partial path-decomposition* $Y^* \in E(Y)$ *with list* $[y^*]$.

Proof. If an element of $y_i^{(u)}$ is repeated, we repeat the corresponding subset $Y_i^{(u)}$ the same number of times. $\square$

Lemma 13.7.2 *Let* Y *be a partial path-decomposition with characteristic* $(Z, \tau[y])$. *Then there exists a partial path-decomposition* Y' *with the same characteristic such that the list* $[y']$ *of* Y' *satisfies* $[y'] \in E(\tau[y])$.

Proof. Assume that no integer sequence $y^{(u)}$ of $[y]$ has two consecutive elements that are the same. Recall the proof of Lemma 13.4.5. Consider a typical operation applied to an integer sequence $y^{(u)}$ of the list $[y]$ of Y. A subsequence $y^{(u)}(i+1 \ldots j-1)$ is replaced by its maximum element $y_k^{(u)}$. We write $Y_j^{(u)}$ for the set of the partial path-decomposition Y corresponding to $y_j^{(u)}$. Consider the path-decomposition Y^* obtained as follows. Initialize $Y_j^* = Y_j$ for all sets Y_j of the path-decomposition Y. Recursively for $j = k-1, \ldots, i+1$ add elements of $Y_{j+1}^{*(u)} \setminus Y_j^{*(u)}$ to the set $Y_j^{*(u)}$ until $Y_j^{*(u)}$ has the same number of elements as $Y_k^{(u)}$. For $j = k+1, \ldots, j-1$ recursively add elements of $Y_{j-1}^{*(u)} \setminus Y_j^{*(u)}$ to the set $Y_j^{*(u)}$ until $Y_j^{*(u)}$ has the same number of elements as $Y_k^{(u)}$. It is easy to see that Y^* is a partial path-decomposition with the same characteristic as Y. By induction on the number of typical operations applied the lemma follows. $\square$

Lemma 13.7.3 *Let* p *be a* **join** *node with children* q *and* r. *Let* A *be a partial path-decomposition rooted at* q *and let* B *be a partial path-decomposition rooted at* r. *Let the restriction of* A *be the same as the restriction of* B. *For each* i *define* $C_i = A_i \cup B_i$. *Then* C *is a partial path-decomposition rooted at* p.

Proof. The restriction of A is a partial path-decomposition for the subgraph induced by X_q. Since p is a **join** node $X_p = X_q = X_r$. Notice that A and B contain the same number of subsets, hence C is well defined. The lemma now follows easily from the definition of a path-decomposition.   □

Definition 13.7.1 *Let p be a join node with children q and r. Let A be a path-decomposition rooted at q and let B be a path-decomposition rooted at r, such that the restrictions of A and B are the same. Then we write $C = A \cup B$ for the path-decomposition rooted at p obtained by $C_i = A_i \cup B_i$ for all i.*

In the next three results we assume FS(p) is computed from full sets of characteristics FS(q) and FS(r) as described in this section.

Theorem 13.7.1 *Let p be a join node with children q and r. For each $(Z, \tau[c]) \in FS(p)$ there is a partial path-decomposition rooted at p with this characteristic.*

Proof. Let A be a partial path-decomposition at q with characteristic $(Z, \tau[a]) \in FS(q)$ and let B be a partial path-decomposition at r with characteristic $(Z, \tau[b]) \in FS(r)$ with the same interval model Z. By Lemma 13.7.2 we may assume that the lists [a] of A and [b] of B satisfy $[a] \in E(\tau[a])$ and $[b] \in E(\tau[b])$. Define

$$[a'] = (a^{(1)} - |Z_{t_1}|, \ldots, a^{(w)} - |Z_{t_w}|) \text{ and}$$
$$[a^*] = (\tau(a^{(1)}) - |Z_{t_1}|, \ldots, \tau(a^{(w)}) - |Z_{t_w}|)$$

Clearly $[a'] \in E[a^*]$ since $[a] \in E(\tau[a])$. Let $[c] \in [a^*] \oplus \tau[b]$ with $\max([c]) \leq \ell + 1$. By Lemma 13.4.14.5 we may conclude that there is a list $[c^\circ] \in E[c]$ such that $[c^\circ] \in [a'] \oplus [b]$. Hence there are extensions $[a^\circ] \in E[a']$ and $[b^\circ] \in E[b]$ such that $[c^\circ] = [a^\circ] + [b^\circ]$. Notice that since $[c^\circ] \in E[c]$ also $\max([c^\circ]) \leq \ell + 1$.

Now take extensions $A^\circ \in E(A)$, corresponding with the extension $[a^\circ]$, and $B^\circ \in E(B)$ corresponding with $[b^\circ]$. Define $C^\circ = A^\circ \cup B^\circ$ (since $[a^\circ]$ and $[b^\circ]$ have the same length in the strong sense C° is well defined). By Lemma 13.7.3 C° is a partial path-decomposition rooted at p. The list of C° is $[c^\circ]$ and hence C° has width at most ℓ. Finally, since $[c^\circ] \in E[c]$: $\tau[c^\circ] = \tau[c]$ (Lemma 13.4.14.2). Hence the characteristic of C° is $(Z, \tau[c]) \in FS(p)$.   □

Theorem 13.7.2 *Let p be a join node with children q and r. If Y is a partial path-decomposition rooted at p of width at most l then there is a partial path-decomposition $Y' \prec Y$ such that $C(Y') \in FS(p)$.*

Proof. Let A be the subdecomposition of Y for G_q and let B be the sub-decomposition of Y for G_r, so $Y = A \cup B$. Since $FS(q)$ and $FS(r)$ are full set of characteristics, we know there exist path-decompositions $A_0 \prec A$ for G_q of which the characteristic is in $FS(q)$ and $B_0 \prec B$ for G_r of which the characteristic is in $FS(r)$. By Lemma 13.7.2 there exists also a partial path-decomposition A' with the same characteristic as A_0, such that $[a'] \in E(\tau[a'])$. Notice that

$$[a'] \equiv \tau[a'] = \tau[a_0] \equiv [a_0] \prec [a]$$

hence $A' \prec A$. In the same manner we find a partial path-decomposition $B' \prec B$ such that $[b'] \in E(\tau[b'])$. Notice that the interval model of all these path-decompositions is the same, say $(Z_{t_i})_{1 \le i \le w}$. Define the list

$$[y^*] = (y^{(1)} + |Z_{t_1}|, \ldots, y^{(w)} + |Z_{t_w}|)$$

(where $[y]$ is the list of Y). Then we have $[y^*] = [a] + [b]$. By Lemma 13.4.14.6 there exists a list $[y^\circ] \in [a'] \oplus [b']$ such that $[y^\circ] \prec [y^*]$. Hence there are extensions $[a^\circ] \in E[a']$ and $[b^\circ] \in E[b']$ such that $[y^\circ] = [a^\circ] + [b^\circ]$. By Lemma 13.7.1 there are path-decompositions $A^\circ \in E(A')$ with list $[a^\circ]$ and $B^\circ \in E(B')$ with list $[b^\circ]$. Define $Y^\dagger = A^\circ \cup B^\circ$. Notice that $Y^\dagger$ is a partial path-decomposition rooted at p with list

$$[y^\dagger] = (y^{\circ(1)} - |Z_{t_1}|, \ldots, y^{\circ(w)} - |Z_{t_w}|)$$

Notice that $[y^\dagger] \prec [y]$ (since $[y^\circ] \prec [y^*]$), hence $Y^\dagger \prec Y$. Since $[a'] \in E(\tau[a'])$ and $[a^\circ] \in E[a']$: $[a^\circ] \in E(\tau[a'])$. Also $[b^\circ] \in E(\tau[b'])$. Hence $[y^\circ] \in \tau[a'] \oplus \tau[b']$. If we define

$$[a'^*] = (\tau(a'^{(1)}) - |Z_{t_1}|, \ldots, \tau(a'^{(w)}) - |Z_{t_w}|)$$

we find $[y^\dagger] \in [a'^*] \oplus \tau[b']$, hence $C(Y^\dagger) \in FS(p)$. $\square$

Corollary 13.7.1 $FS(p)$ *is a full set of characteristics for the **join** node* p.

13.8 A full set for a forget node

Let p be a **forget** node with child q. Then $G_p = G_q$ and by Definitions 13.1.4 and 13.1.5 $X_p \subset X_q$ and X_q contains exactly one vertex, say x, which is not in X_p. We call x the *forgotten* element of p. We first show how to compute the full set of characteristics $FS(p)$ from the full set of characteristics $FS(q)$.

Let $(Z, \tau[y])$ be a characteristic in $FS(q)$, with interval model $Z = (Z_{t_j})_{1 \le j \le w}$. Since Z is a path-decomposition for the subgraph induced by X_q there is a consecutive number of subsets in Z which contain the forgotten element x. We remove x from these sets, and remove consecutive subsets which are now the same. In this way we obtain an interval model Z' for p.

Lemma 13.8.1 Z' *is an interval model for* p.

Proof. Obvious.   □

Let $i \leq j$ be such that Z_{t_i} is the first subset Z which contains x and Z_{t_j} is the last subset containing x. Notice that the number of subsets in Z is at most two more than the number of subsets of Z', namely Z_{t_i} can become the same as $Z_{t_{i-1}}$ after the removal of x and Z_{t_j} can become the same as $Z_{t_{j+1}}$. Consider the following four cases.

1. If the number of subsets of Z' is the same as the number of subsets in Z, then we put $(Z', \tau[y])$ in FS(p).

2. If only the subset $Z_{t_i} \setminus \{x\}$ is the same as $Z_{t_{i-1}}$ then let

$$\tau^* = \tau(\tau(y^{(i-1)}) +\!\!+ \tau(y^{(i)}))$$

   and change the typical list $\tau[y]$ into the list:

$$[y'] = (\tau(y^{(1)}), \dots, \tau(y^{(i-2)}), \tau^*, \tau(y^{(i+1)}), \dots, \tau(y^{(w)}))$$

   i.e. we concatenate the typical sequences $\tau(y^{(i-1)})$ and $\tau(y^{(i)})$ and compute the typical sequence of the result. We put $(Z', [y'])$ in FS(p).

3. If only $Z_{t_j} \setminus \{x\} = Z_{t_{j+1}}$ then compute

$$\tau(\tau(y^{(j)}) +\!\!+ \tau(y^{(j+1)}))$$

   and change the typical list $\tau[y]$ into $[y'']$ as in the former case. Put $(Z', [y''])$ into FS(p).

4. Finally, if both $Z_{t_{i-1}} = Z_{t_i} \setminus \{x\}$ and $Z_{t_j} \setminus \{x\} = Z_{t_{j+1}}$ then let

$$\tau_1 = \tau(\tau(y^{(i-1)}) +\!\!+ \tau(y^{(i)})) \wedge \tau_2 = \tau(\tau(y^{(j)}) +\!\!+ \tau(y^{(j+1)}))$$

   and change the typical list $\tau[y]$ into the list:

$$[y^*] = (\tau(y^{(1)}), \dots, \tau(y^{(i-2)}), \tau_1, \tau(y^{(i+1)}), \dots$$
$$\dots, \tau(y^{(j-1)}), \tau_2, \tau(y^{(j+2)}), \dots, \tau(y^{(w)}))$$

   We put $(Z', [y^*])$ in FS(p).

Notice that if $i = j$ we compute in this last case the typical sequence of:
$\tau(y^{(i-1)}) +\!\!+ \tau(y^{(i)}) +\!\!+ \tau(y^{(i+1)})$.

FS(p) is obtained by carrying out the above for each element of FS(q). Below we assume FS(p) is computed in this way from FS(q). To prove the correctness we first show that an element of FS(p) is a characteristic of a partial path-decomposition rooted at p.

Theorem 13.8.1 *For each* $(Z', [c]) \in FS(p)$ *there is a partial path-decomposition rooted at* p *with this characteristic.*

Proof. Let $(Z, \tau[y])$ be the corresponding characteristic in $FS(q)$ (i.e. $(Z', [c])$ is computed from $(Z, \tau[y])$ by the algorithm described above). There exists a partial path-decomposition Y rooted at q, with this characteristic. Y is also a partial path-decomposition rooted at p. By Lemma 13.8.1 the interval model of Y at node p is Z'. We prove that the typical list is computed correctly. This is clearly the case when $Z' = Z$. Consider the second case: the number of subsets in Z' is one less and $Z_{t_{i-1}} = Z_{t_i} \setminus \{x\}$. Let $[y]$ be the list of Y. In this case the list of Y changes into

$$(y^{(1)}, \ldots, y^{(i-2)}, y^{(i-1)} + \! + y^{(i)}, y^{(i+1)}, \ldots, y^{(w)})$$

By Lemma 13.4.10: $\tau(y^{(i-1)} + \! + y^{(i)}) = \tau(\tau(y^{(i-1)}) + \! + \tau(y^{(i)}))$, hence the typical list is computed correctly. The other cases are similar. □

The next theorem shows that $FS(p)$ is indeed a full set of characteristics.

Theorem 13.8.2 *If* Y *is a partial path-decomposition rooted at* p *with width at most* ℓ, *then there is a partial path-decomposition* $Y' \prec Y$ *such that* $C(Y') \in FS(p)$.

Proof. Y is also a partial path-decomposition rooted at q, since $G_q = G_p$. Hence there is a partial path-decomposition rooted at q, $Y' \prec Y$, of which the characteristic is in $FS(q)$. In the proof of Theorem 13.8.1 it is shown that the characteristic of Y' is computed correctly for node p. We only have to show that $Y' \prec Y$ still holds for node p (recall Definition 13.5.5: the interval model may have changed!). This however is proved in Lemma 13.4.12. □

Corollary 13.8.1 $FS(p)$ *is a full set of characteristics for the* **forget** *node* p.

13.9 A full set for an introduce node

In this last subsection we consider the case in which p is an **introduce** node with child q. In this case $V_p \setminus \{x\} = V_q$, for some vertex x, where V_i is the vertex set for subgraph G_i. We call the vertex x *introduced* at p. The neighbors of x are all contained in X_p. We first give an algorithm to compute a full set of characteristics $FS(p)$ for p.

To simplify the presentation we compute $FS(p)$ as follows. We first make a list of all feasible interval models for the for the node p (a minimal path-decomposition for the subgraph induced by X_p is a feasible interval model for p). We then check, for each feasible interval model, if there is a characteristic in $FS(q)$ which can be 'extended' to a characteristic for X_p with this interval model. Clearly, this algorithm might not be the most efficient one to compute $FS(p)$, since there could be many feasible interval models which are in fact not interval models.

Algorithm

Step 1 Make a list Q of all minimal path-decompositions for the subgraph induced by X_p (Definition 13.3.3).

Step 2 Make a new list Q^* as follows. For each minimal path-decomposition $Z \in Q$ compute Z': Remove the introduced vertex x from all subsets of Z and after that remove repetitions of subsets. Notice that Z' is a minimal path-decomposition for the subgraph induced by X_q. Put the pair (Z, Z') in Q^*.

Step 3 If for some pair $(Z, Z') \in Q^*$ there is *no* characteristic in the full set of characteristics for q, $FS(q)$, which has Z' as an interval model, then remove the pair (Z, Z') from the list Q^*.

Step 4 Initialize $FS(p) = \emptyset$. For each pair (Z, Z') in Q^* and for each characteristic $(Z', \tau[c]) \in FS(q)$ with Z' as an interval model do the following. Let $Z = (Z_{t_s})_{1 \le s \le w}$. Let $i \le j$ be such that Z_{t_i} is the first subset of Z containing the introduced vertex x and Z_{t_j} the last subset of Z containing x. Notice that the number of subsets of Z is at most two more than the number of subsets of Z'. Namely, after the removal of x Z_{t_i} can become the same as $Z_{t_{i-1}}$ and Z_{t_j} can become the same as $Z_{t_{j+1}}$. Hence one of the following four different cases is applicable:

1. If the number of subsets of Z' is the same as the number of subsets in Z, we change the typical list $\tau[c]$ into:
$$[c^\circ] = (\tau(c^{(1)}), \ldots, \tau(c^{(i-1)}), 1 + \tau(c^{(i)}), \ldots$$
$$\ldots, 1 + \tau(c^{(j)}), \tau(c^{(j+1)}), \ldots, \tau(c^{(w)}))$$
   i.e. we add one to all typical sequences $\tau(c^{(u)})$ with $i \le u \le j$. If $\max([c^\circ]) \le \ell + 1$, then we put $(Z, [c^\circ])$ in $FS(p)$.

2. If the number of subsets in Z' is one less than the number of subsets in Z, and $Z_{t_i} \setminus \{x\} = Z_{t_{i-1}}$: For convenience we write:
$$\tau[c] = (\tau(c^{(1)}), \ldots, \tau(c^{(i-2)}), \tau(c^{(i)}), \tau(c^{(i+1)}), \ldots, \tau(c^{(w)}))$$
   Consider *all splits of both types* (δ_1, δ_2) of $\tau(c^{(i)})$ (Definition 13.4.6). For each such split change the typical list $\tau[c]$ into:
$$[c'] = (\tau(c^{(1)}), \ldots, \tau(c^{(i-2)}), \delta_1, 1 + \delta_2, 1 + \tau(c^{(i+1)}), \ldots$$
$$\ldots, 1 + \tau(c^{(j)}), \tau(c^{(j+1)}), \ldots, \tau(c^{(w)}))$$
   If $\max([c']) \le \ell + 1$ then we put $(Z, [c'])$ in $FS(p)$.

3. If the number of subsets in Z' is one less than the number of subsets in Z, and $Z_{t_j} \setminus \{x\} = Z_{t_{j+1}}$: This case is similar to the second case. In this case we make all splits of $\tau(c^{(j)})$.

4. If the number of subsets in Z' is two less than the number of subsets in Z: Let

$$\tau[c] = (\tau(c^{(1)}), \ldots, \tau(c^{(i-2)}), \tau(c^{(i)}), \ldots, \tau(c^{(j)}), \tau(c^{(j+2)}), \ldots, \tau(c^{(w)}))$$

In this case consider all splits (α_1, α_2) of $\tau(c^{(i)})$ and all splits (β_1, β_2) of $\tau(c^{(j)})$. For each pair of such splits (α_1, α_2) and (β_1, β_2) change the typical list $\tau[c]$ into

$$[c^\dagger] = (\tau(c^{(1)}), \ldots, \tau(c^{(i-2)}), \alpha_1, 1 + \alpha_2, 1 + \tau(c^{(i+1)}), \ldots$$
$$\ldots, 1 + \beta_1, \beta_2, \tau(c^{(j+2)}), \ldots, \tau(c^{(w)}))$$

If $\max([c^\dagger]) \le \ell + 1$ then put $(Z, [c^\dagger]$ in FS(p).

Notice that in the last case, if $i = j$ then we split $\tau(c^{(i)})$ into *three* parts; i.e first split it into two parts and then split the second part again.

Step 5 Stop. The computation of FS(p) is completed.

We prove that FS(p) is a full set of characteristics for p in two stages. First we demonstrate that every element in FS(p) is a characteristic of a partial path-decomposition rooted at p.

Theorem 13.9.1 *For each* $(Z, [d]) \in$ FS(p) *there is a partial path-decomposition rooted at* p *with this characteristic.*

Proof. Let $(Z', \tau[y])$ be the corresponding characteristic in FS(q), i.e. $(Z, [d])$ is computed from this characteristic by the algorithm described above. There is a partial path-decomposition Y rooted at q, with $C(Y) = (Z', \tau[y])$. Let $[y]$ be the list for Y. By Lemma 13.7.2 we may assume that $[y] \in E(\tau[y])$. Let $(Z', [Y])$ be the list representation of Y.

First consider the case with $|Z'| = |Z|$ (the same number of subsets). If $i \le j$ are the first and last subset of Z containing x, we change the path-decomposition Y into Y°, by adding x to all subsets of $Y^{(u)}$, for all $i \le u \le j$. Clearly, Y° is a partial path-decomposition for G_p. Since $\tau(y^{\circ(u)}) = \tau(1 + y^{(u)}) = 1 + \tau(y^{(u)})$ for all $i \le u \le j$, the typical list $\tau[y^\circ]$ for Y° equals the list $[d]$ as computed by the algorithmn. Hence Y° is a partial path-decomposition with characteristic $(Z, [d])$.

Now consider the case where the number of subsets in Z is one more than the number of subsets in Z' and $Z_{t_i} \setminus \{x\} = Z_{t_{i-1}}$ (the second case). In this case the typical sequence $\tau(y^{(i)})$ is split by the algorithm into, say,

(δ_1, δ_2) in order to obtain $[d]$. The integer sequence $y^{(i)}$ is an extension of $\tau(y^{(i)})$. We make a split of $y^{(i)}$, say $(y_1^{(i)}, y_2^{(i)})$ such that $y_1^{(i)} \in E(\delta_1)$ and $y_2^{(i)} \in E(\delta_2)$. By Lemma 13.4.13 this split is always possible and $\tau(y_1^{(i)}) = \delta_1$ and $\tau(y_2^{(i)}) = \delta_2$. Take a similar 'split' of $Y^{(i)}$ into $(Y_1^{(i)}, Y_2^{(i)})$. Add the vertex x to all subsets of $Y_2^{(i)}$, and to all subsets of $Y^{(u)}$ with $i+1 \leq u \leq j$. Call the obtained path-decomposition Y° (notice that Y° is indeed a partial path-decomposition rooted at p of width at most ℓ). Since the typical sequence for $y_1^{(i)}$ is δ_1 and the typical sequence for $y_2^{(i)}$ is δ_2, the characteristic of Y° is indeed $(Z, [d])$, where $[d]$ is the typical list of Y° as computed by the algorithm.

The other two cases are similar. $\qquad\square$

Theorem 13.9.2 *For every partial path-decomposition Y rooted at p of width at most ℓ there exists a partial path-decomposition $Y' \prec Y$, such that $C(Y') \in FS(p)$.*

Proof. Let Y° be the subdecomposition of Y for G_q. Since $FS(q)$ is a full set of characteristics, there exists a partial path-decomposition $Y_0 \prec Y^\circ$ of which the characteristic is in $FS(q)$. By Lemma 13.7.2 we know there exists a partial path-decomposition Y' with the same characteristic as Y_0, such that $[y'] \in E(\tau[y'])$ (where $[y']$ is the list of Y' and $\tau[y']$ is the typical list for Y'). Notice that $[y'] \equiv \tau[y'] = \tau[y_0] \equiv [y_0] \prec [y^\circ]$ (Lemma 13.4.14 and Definition 13.5.5) hence $Y' \prec Y^\circ$. So there are extensions $Y'^* \in E(Y')$ and $Y^{\circ *} \in E(Y^\circ)$ such that the respective lists satisfy $[y'^*] \leq [y^{\circ *}]$ (Lemma 13.7.1). Since Y and Y° have the same length we can take an extension $Y^* \in E(Y)$ corresponding with $Y^{\circ *}$ (i.e. if some subset Y_f° is repeated r times in $Y^{\circ *}$ then we repeat Y_f also r times in Y^*). Notice that now we have three partial path-decompositions of the same length Y^*, $Y^{\circ *}$ and Y'^* and that $Y^{\circ *}$ and Y'^* have the same interval model.

Make a partial path-decomposition $Y^\dagger$ rooted at p by changing Y'^* as follows. Add x to $Y_f'^*$ whenever $x \in Y_f^*$. *Notice that the interval model of $Y^\dagger$ is the same as the interval model of Y* and $[y^\dagger] \leq [y^*]$, hence $Y^\dagger \prec Y$.

We now show that the characteristic of $Y^\dagger$ is in $FS(p)$. Clearly, the characteristic of Y'^* is in $FS(q)$. Let $(Z', [Y'^*])$ be the list representation for Y'^* and let $(Z, [Y^\dagger])$ be the list representation for $Y^\dagger$. Write $Z = (Z_{t_s})_{1 \leq s \leq w}$. Let $i \leq j$ be the first and last interval of Z containing the vertex x.

Consider the case where the number of intervals of Z is the same as the number of intervals of Z'. Then x is added to *all* subsets of $Y'^{*(u)}$ for all $i \leq u \leq j$ (otherwise at least one interval of Z' would have been split). It follows that the characteristic of $Y^\dagger$ is computed in the first case of step 4 of the algorithm: $y^{\dagger(u)} = 1 + y'^{*(u)}$ for all $i \leq u \leq j$ hence $\tau(y^{\dagger(u)}) = 1 + \tau(y'^{*(u)})$.

Next consider the case where the number of intervals of Z' is one less than the number of intervals of Z and $Z_{t_i} \setminus \{x\} = Z_{t_{i-1}}$. For convenience we write:

$$[y'^*] = (y'^{*(1)}, \ldots, y'^{*(i-2)}, y'^{*(i)}, \ldots, y'^{*(j)}, y'^{*(j+1)}, \ldots, y'^{*(w)})$$

So $y'^{*(i)} = y^{\dagger(i-1)} ++(y^{\dagger(i)} - 1)$. Now $[y'^*] \in E(\tau[y'])$. Write $\tau(y'^{(i)}) = (\alpha_1, \ldots, \alpha_s)$. Then it follows that either

$$
\begin{aligned}
y^{\dagger(i-1)} &= (\alpha_1, \ldots, \alpha_1, \ldots, \alpha_f, \ldots, \alpha_f) \text{ and} \\
y^{\dagger(i)} &= 1 + (\alpha_f, \ldots, \alpha_f, \ldots, \alpha_s, \ldots, \alpha_s) \text{ or} \\
y^{\dagger(i-1)} &= (\alpha_1, \ldots, \alpha_1, \ldots, \alpha_f, \ldots, \alpha_f) \text{ and} \\
y^{\dagger(i)} &= 1 + (\alpha_{f+1}, \ldots, \alpha_{f+1}, \ldots, \alpha_s, \ldots, \alpha_s)
\end{aligned}
$$

It follows that the characteristic of $Y^{\dagger}$ is computed by the algorithm in the second case of step 4.

The other cases are similar. $\square$

Corollary 13.9.1 *The set* $FS(p)$ *computed by the algorithm is a full set of characteristics for the* **introduce** *node* p.

Chapter 14

Treewidth of treewidth-bounded graphs

In this chapter we give a linear time algorithm to determine the treewidth of a graph, if the graph has bounded treewidth. We assume an approximate tree-decomposition of width bounded by a constant is given. The algorithm is based on the results of the previous chapter. The method used to find the pathwidth of a graph can be extended to find the treewidth. We show that a tree-decomposition of width bounded by some constant can be characterized using a description of constant size. This description resembles the characteristic of a path-decomposition of the previous chapter. In fact, the characteristic tree-decomposition we define in this chapter, has, as one ingredient, characteristic path-decompositions of parts of the tree-decomposition. The tree itself is characterized by a concept we call the trunk. Recently, using our algorithm as a subroutine, Bodlaender succeeded in developing a linear time algorithm for treewidth if the treewidth is bounded by some constant. This algorithm improves many earlier results [4, 57, 115, 139]. In the next chapter we describe this algorithm.

14.1 Preliminaries

We assume we have a nice tree-decomposition $NT = (\{X_i | i \in I\}, T = (I, F))$ of width k of the input graph $G = (V, E)$. In this section we give a linear time algorithm to decide whether the treewidth of G is at most ℓ (k and ℓ are assumed constant). We start by defining in this section the characteristic of a tree-decomposition. We may restrict ourselves to tree-decompositions which are minimal in some sense.

Definition 14.1.1 Let $D = (S, T)$ be a tree-decomposition for a graph G. D is called non-trivial if for every pair of adjacent nodes x and y in T the corresponding subsets S_x and S_y are different.

Clearly, if D is a tree-decomposition, it can be transformed to a tree-decomposition which is non-trivial.

Definition 14.1.2 *Let* $D = (S, T)$ *be a tree-decomposition. Let* x *be a leaf of* T *and let* S_x *be the corresponding subset. The leaf* x *is called* maximal *if* S_x *contains a vertex* v *which is not an element of any other subset of* S.

Notice that if x is a leaf and if y is the father of x, then x is exactly maximal if S_x is *not* a subset of S_y.

Definition 14.1.3 *Let* $D = (S, T)$ *be a tree-decomposition for a graph* G. D *is called* minimal *if the following two conditions are satisfied:*

1. D *is non-trivial, and*

2. *all leaves of* T *are maximal.*

Lemma 14.1.1 *Let* G *be a graph with treewidth* ℓ. *There exists a minimal tree-decomposition of* G *of width* ℓ.

Proof. Take any tree-decomposition $D = (S, T)$ of G of width ℓ. We transform D into a minimal tree-decomposition D' as follows. First, recursively remove leaves of T which are not maximal and remove the corresponding subsets from S. If $e = (x, y)$ is an edge of T such that $S_x = S_y$, then contract the edge in T and replace the subsets S_x and S_y by one new subset. It is easy to see that D' obtained in this way is a minimal tree-decomposition. □

We want to show an upper bound on the number of nodes in a minimal tree-decomposition. The following lemma will be useful.

Lemma 14.1.2 *Let* T *be a tree with* $p \geq 2$ *nodes* b *leaves. There are at most* $b - 2$ *nodes in* T *with degree at least three. Equality holds if and only if the maximum degree is at most three.*

Proof. For each vertex x of T let d_x be the degree. Count the number of edges in T:

$$\begin{aligned}
2(p - 1) &= \sum_{x, d_x = 1} 1 + 2 \sum_{x, d_x = 2} 1 + 3 \sum_{x, d_x = 3} 1 + 4 \sum_{x, d_x = 4} 1 + \ldots \\
&= 2p - b + \sum_{x, d_x = 3} 1 + 2 \sum_{x, d_x = 4} 1 \ldots \\
&\geq 2p - b + \sum_{x, d_x \geq 3} 1
\end{aligned}$$

Hence $\sum_{x, d_x \geq 3} 1 \leq b - 2$. □

Lemma 14.1.3 *Let* G *be a graph with* n *vertices. Then the number of nodes in a minimal tree-decomposition is at most* $(2n-1)^2$.

Proof. Let $D = (S, T)$ be a minimal tree-decomposition for a graph G with n vertices. Let H be the chordal completion of G implied by D (Definition 2.2.6 on page 19). Since each leaf is maximal, the corresponding subsets are maximal cliques in H and all are different. Since H is triangulated, the number of maximal cliques in H is at most n (see for example [66]). Hence, T has at most n leaves, and by Lemma 14.1.2 the number of vertices of degree at least three is at most $n - 1$ (we add one to account for the case $n = 1$). Since adjacent subsets are different, we can use Lemma 13.3.1 (page 152) to see that each path in T with all vertices of degree two can have length at most $2n - 1$. It follows that T has at most $(2n - 2)(2n - 1)$ vertices of degree two. Hence, the total number of nodes in T is at most $(2n-1)^2$. $\square$

Recall that we have a nice tree-decomposition $NT = (\{X_i | i \in I\}, T = (I, F))$. Recall Definition 13.1.2 (page 148) of a rooted subgraph.

Definition 14.1.4 *A partial tree-decomposition rooted at a node* $i \in I$ *is a tree-decomposition for* G_i, *i.e., the subgraph rooted at* i.

Definition 14.1.5 *Let* $Y = (SY, TY)$ *be a partial tree-decomposition rooted at a node* i. *The* restriction *of* Y *is the sub-decomposition* $Y^* = (SY^*, TY)$ *for the subgraph induced by* X_i, *i.e. if we define for each* $S \in SY$, $S' = S \cap X_i$, *then* $SY^* = \{S' \mid S \in SY\}$.

The characteristic of a partial tree-decomposition consists of three parts. We call the first part the trunk of the tree-decomposition.

Definition 14.1.6 *Let* $Y = (SY, TY)$ *be a partial tree-decomposition rooted at a node* i. *The* trunk *of* Y *is a tree* T *defined as follows. First take the restriction of* Y, *say* $Y^* = (SY^*, TY)$. *Next, recursively remove leaves of* TY *for which the corresponding subsets of* SY^* *are not maximal in* SY^*. *Finally, remove those vertices of the tree which have degree two and make the two neighbors adjacent.*

Lemma 14.1.4 *Let* Y *be a partial tree-decomposition rooted at a node* i. *The number of vertices of the trunk of* Y *is at most* $2k$.

Proof. The number of vertices of X_i is at most $k + 1$. Hence the number of maximal subsets of SY^* is at most $k + 1$ (see the proof of Lemma 14.1.3). By Lemma 14.1.2, the number of nodes with degree at least three in the trunk is at most $k - 1$. Since each node in the trunk is either a leaf or a node with degree at least three, the number of nodes is bounded by $2k$. $\square$

We now define the tree model of a partial tree-decomposition in analogue of the interval model defined in Definition 13.3.2. Let $Y = (SY, TY)$ be a partial tree-decomposition roote at a node i and let T be the trunk. For each edge e in the trunk, consider the corresponding path in TY with all internal vertices of degree 2. Let SY_e be the corresponding subsets of SY. We use the notation Y_e for the pair (SY_e, TY_e). Notice that Y_e is a path-decomposition for the subgraph induced by the vertices in $\cup_{S \in SY_e} S$.

Definition 14.1.7 *Let* $Y = (SY, TY)$ *be a partial tree-decomposition rooted at a node* i. *The* tree model *of* Y *is a pair*

$$(T, (Z_e)_{e \in T})$$

where T *is the trunk of* Y *and* Z_e *is the interval model of* Y_e *for each edge* e *of* T.

Recall Definition 13.5.1 (page 161) of the list representation of of a path-decomposition.

Definition 14.1.8 *Let* Y *be a partial tree-decomposition rooted at a node* i. *The* trunk-representation *is:*

$$(T, (Z_e)_{e \in T}, ([Y_e])_{e \in T})$$

where $(Z_e, [Y_e])$ *is the list representation for* Y_e *(for each edge* e *in the trunk).*

Recall Definition 13.5.3 (page 162) of a typical list of a path-decomposition.

Definition 14.1.9 *Let* Y *be a partial tree-decomposition rooted at a node* i *with tree model* $(T, (Z_e)_{e \in T})$. *The* characteristic *of* Y *is the triple:*

$$C(Y) = (T, (Z_e)_{e \in T}, (\tau[y_e])_{e \in T})$$

where $\tau[y_e]$ *is the typical list of* Y_e.

Notice that the characteristic is of constant size by Lemma 14.1.4 and Lemma 13.5.1 (page 162).

Definition 14.1.10 *Let* Y *and* Z *be two partial tree-decompositions rooted at the same node* i, *which have the same tree model (i.e., the same trunk and for each edge of the trunk the same interval model). Then* $Y \prec Z$ *if for each edge* e *in the trunk the corresponding lists satisfy* $[y_e] \prec [z_e]$. *If* $Y \prec Z$ *and* $Z \prec Y$ *then we write* $Y \equiv Z$.

Recall Definition 13.5.6 (page 162) for the full set of characteristics. We define the full set of characteristics for partial tree-decompositions.

Definition 14.1.11 *A set of characteristics* FS(i) *of partial tree-decompositions rooted at some node* i *of width at most* ℓ *is called a* full set of characteristics *if for each partial tree-decomposition* Y *rooted at* i *and of width at most* ℓ *either its characteristic is in* FS(i) *or there is a partial tree-decomposition* Y' *with* $Y' \prec Y$ *and the characteristic of* Y' *is in* FS(i).

In the following sections we show how to compute a full set of characteristics for each node from the full sets of characteristics of the children of the node.

14.2 A full set for a start node

Let p be a **start** node. Then G_p is the subgraph induced by X_p. We show to compute the full set FS(p) of characteristics. Generate all minimal tree-decompositions $Y = (SY, TY)$ for the subgraph induced by X_p of width at most ℓ. This can be done in constant time by Lemma 14.1.3. The trunk T of Y in this case can be obtained by removing nodes of TY which have degree two. For each edge $e \in T$, the interval model Z_e is simply the sequence $[Y_e]$ (since Y is minimal). The typical sequence for the ιth interval satisfies $\tau(y_e^{(\iota)}) = (|Y_e^{(\iota)}|)$, i.e. consists of one element. In other words $\tau[y_e] = [y_e]$.

Lemma 14.2.1 *If* p *is a* **start** *node, then* FS(p) *is a full set of characteristics.*

Proof. Let Y be a partial tree-decomposition rooted at p, of width at most ℓ. Notice that if we remove leaves which are not maximal, this does not change the characteristic. Also, if two adjacent subsets are equal, we can contract the edge without changing the characteristic. This shows that for every partial tree-decomposition with width at most ℓ, the characteristic is in FS(p). $\square$

14.3 A full set for a join node

Let p be a **join** node with children q and r. By definition $X_p = X_q = X_r$. First we show how to compute a full set of characteristics FS(p) from the sets FS(q) and FS(r). Take $(T, (Z_e)_{e\in T}, (\tau[a_e])_{e\in T}) \in FS(q)$ and $(T, (Z_e)_{e\in T}, (\tau[b_e])_{e\in T}) \in FS(r)$. Hence these characteristics have *the same tree model*. Compute, for each edge e of the trunk, the list

$$[a_e^*] = (\tau(a_e^{(1)}) - |Z_e^{(1)}|, \tau(a_e^{(2)}) - |Z_e^{(2)}|, \ldots) \qquad (14.1)$$

Then compute *all* list $\tau[c_e]$, where $[c_e] \in [a_e^*] \oplus \tau[b_e]$ with $\max([c_e]) \le \ell+1$. Put

$$(\mathcal{T}, (Z_e)_{e \in \mathcal{T}}, (\tau[c_e])_{e \in \mathcal{T}})$$

in the set $FS(p)$ for all possible choices of $\tau[c_e]$.

In proving the correctness of the above construction, the following notion of the join of two partial tree-decompositions will be useful. Let p be a **join** node with children q and r. Let $A = (SA, TA)$ be a partial tree-decomposition rooted at q and let $B = (SB, TB)$ be a partial tree-decomposition rooted at r. Let $A' = (SA', TA)$ be the restriction of A and let $B' = (SB', TB)$ be the restriction of B. From the trees TA and TB recursively remove leaves which are not maximal in the restriction. Call these new trees TA^* and TB^*, and let SA^* and SB^* be those subsets of SA' and SB' corresponding with nodes in TA^* and TB^* respectively. Assume that:

$$(SA^*, TA^*) = (SB^*, TB^*)$$

We define a tree-decomposition C rooted at p as follows. Define a tree TC by taking the union of TA and TB and by *identifying* the nodes of TA^* and TB^*. For each node x of TC, define a subset SC_x as:

$$SC_x = \begin{cases} SA_x \cup SB_x & \text{if } x \in TA^* \\ SA_x & \text{if } x \in TA \setminus TA^* \\ SB_x & \text{if } x \in TB \setminus TB^* \end{cases}$$

If A and B are partial tree-decompositions rooted at q and r, satisfying the conditions, then we write $C = A \cup B$ for the tree-decomposition defined above.

Remark. Notice that the trunks of the three partial tree-decompositions A, B and C are the same.

Lemma 14.3.1 $C = A \cup B$ *is a partial tree-decomposition rooted at* p.

Proof. The conditions are easily verified. $\qquad\qquad\qquad\qquad\qquad\qquad\square$

Theorem 14.3.1 *Let* p *be a* **join** *node with children* q *and* r. *Let*

$$C = (\mathcal{T}, (Z_e)_{e \in \mathcal{T}}, (\tau[y_e])_{e \in \mathcal{T}}) \in FS(p)$$

There exists a partial tree-decomposition of width at most ℓ *and rooted at* p *with this characteristic.*

Proof. Let

$$C(A) = (\mathcal{T}, (Z_e)_{e\in\mathcal{T}}, (\tau[a_e])_{e\in\mathcal{T}}) \in FS(q)$$
$$C(B) = (\mathcal{T}, (Z_e)_{e\in\mathcal{T}}, (\tau[b_e])_{e\in\mathcal{T}}) \in FS(r)$$

Let C be obtained from C(A) and C(B) by the algorithm:

$$\forall_{e\in\mathcal{T}} [[y_e] \in [a_e^*] \oplus \tau[b_e] \wedge \max([y_e]) \leq \ell + 1]$$

where $[a_e^*]$ as defined in formula 14.1. It is easy to verify that the proof of Theorem 13.7.1 generalizes to obtain the following result. There exist partial tree-decompositions A° rooted at q and B° rooted at r with characteristics C(A) and C(B) respectively, such that $C^\circ = A^\circ \cup B^\circ$ is well defined and has characteristic C. □

Theorem 14.3.2 *Let p be a **join** node with children q and r. If Y is a partial tree-decomposition rooted at p of width at mist ℓ then there exists a partial tree-decomposition Y' such that $Y' \prec Y$ and such that $C(Y') \in FS(p)$.*

Proof. Let A be the subdecomposition of Y for G_q and let B be the subdecomposition for G_r. Notice that $\bar{Y} = A \cup B$ is well defined and satisfies $C(\bar{Y}) = C(Y)$. The result of Theorem 13.7.2 can easily be generalized and we obtain the following result. There exist partial tree-decompositions A° rooted at q and B° rooted at r, such that $C(A^\circ) \in FS(q)$, $C(B^\circ) \in FS(r)$ and $Y^\dagger = A^\circ \cup B^\circ$ is well defined and satisfies $Y^\dagger \prec \bar{Y}$. Moreover we may assume that for each edge of the trunk $[a_e^\circ] \in E(\tau[a_e^\circ])$ and $[b_e^\circ] \in E(\tau[b_e^\circ])$. Hence $[y_e^\dagger] \in [a_e^{\circ*}] \oplus \tau[b_e^\circ]$, with $[a_e^{\circ*}]$ computed from $\tau[a_e^\circ]$ as in formula 14.1. This proves the theorem. □

Corollary 14.3.1 $FS(p)$ *is a full set of characteristics for the **join** node p.*

14.4   A full set for a forget node

Let p be a forget node with child q, and let x be the forgotten element i.e. $X_p = X_q \setminus \{x\}$. We first show how to compute a full set of characteristics FS(p) from a full set of characteristics FS(q) for q.

Let $(\mathcal{T}, (Z_e)_{e\in\mathcal{T}}, (\tau[y_e])_{e\in\mathcal{T}})$ be a characteristic of a partial tree-decomposition Y rooted at q. Remove from all sets $Z_e^{(i)}$ the vertex x. Compute the new trunk $\mathcal{T}^*$. Remove repetitions from the interval models Z_e and for each edge $e \in \mathcal{T}^*$ let Z_e^* be the new interval model. Finally, for each edge $e \in \mathcal{T}^*$ change the typical list $\tau[y_e]$ into $\tau[y_e^*]$ as described in section 13.8. Put $(\mathcal{T}^*, (Z_e^*)_{e\in\mathcal{T}^*}, (\tau[y_e^*])_{e\in\mathcal{T}^*})$ in FS(p).

Lemma 14.4.1 *The trunk T^* differs from T, only if there is exactly one subset $Z_e^{(i)}$ containing the vertex x. Moreover, in that case at least one end vertex of e must be a leaf of T.*

Proof. Let $Y^* = (SY^*, TY)$ be the restriction of Y at the node q. Recursively remove leaves from TY which are not maximal, and let TY^* be the new tree. Notice that the trunk T^* can only differ from T if some leaf node w of TY^* is not maximal any more, when x is removed from all subsets. Since the subset corresponding with w is maximal, it contains a vertex z which is not in the subset corresponding with the neighbor of w. Since the only vertex which is removed is the forgotten vertex x, it follows that $z = x$. □

Lemma 14.4.2 *The tree model for Y at p is $(T^*, (Z_e)_{e \in T^*})$.*

Proof. See the proof of Lemma 13.8.1. □

Theorem 14.4.1 *For each $(T^*, (Z_e^*)_{e \in T^*}, (\tau[y_e^*])_{e \in T^*}) \in FS(p)$ there is a partial tree-decomposition Y rooted at p with this characteristic.*

Proof. Let $(T, (Z_e)_{e \in T}, (\tau[y_e])_{e \in T})$ be the corresponding characteristic in FS(q). Then there is a partial tree-decomposition Y rooted at q with this characteristic. Then Y is also a partial tree-decomposition rooted at p since $G_p = G_q$. By Lemma 14.4.2, $(T^*, (Z_e)_{e \in T^*})$ is the tree model for Y at p. By Theorem 13.8.1 for each edge e in T^* the typical list $\tau[y_e^*]$ is computed correctly. □

Theorem 14.4.2 *If Y is a partial tree-decomposition rooted at p of width at most ℓ then there exists a partial tree-decomposition $Y' \prec Y$ such that $C(Y') \in FS(p)$.*

Proof. Y is also a partial tree-decomposition rooted at q. Since FS(q) is a full set of characteristics, there is a partial tree-decomposition Y' with $Y' \prec Y$ such that $C(Y') \in FS(q)$. It is shown that the characteristic of Y' is computed correctly for node p. We have to show that $Y' \prec Y$ holds for node p. Since Y and Y' have the same tree model at q, they also have the same tree model at p (see Lemma 14.4.2; the tree model at p is computed from the tree model at q). Using Lemma 13.4.12 it follows that for each edge $e \in T^*$ we have $[y_e'] \prec [y_e]$. Hence $Y' \prec Y$ for node p. □

Corollary 14.4.1 FS(p) *is a full set of characteristics for the forget node p.*

14.5   A full set for an introduce node

Let p be an **introduce** node with child q. Let x be the vertex introduced at p, i.e. $X_p = X_q \cup \{x\}$. We start by giving the algorithm to compute the full set $FS(p)$. For reasons of simplicity we apply the same method as in section 13.9: First we compute all minimal tree-decompositions for X_p. Notice that by Lemma 14.1.3 this can be done in constant time. Remove the vertex x from all subsets, obtaining a tree-decomposition for X_q. Next compute the tree model of this tree-decomposition. Check if there is a characteristic in $FS(q)$ with this tree model, and change this into a new characteristic for $FS(p)$.

Algorithm

Step 1 Make a list Q of all tree models of minimal tree-decompositions for the graph induced by X_p of width at most ℓ.

Step 2 Make a new list Q' as follows. For each element T^* of Q remove x from all subsets. Compute the tree model $T = (\mathcal{T}, (Z_e)_{e \in \mathcal{T}})$ of the result. If there is a characteristic in $FS(q)$ with this tree model, then put the pair (T^*, T) in the list Q'.

Step 3 Let $(T^*, T) \in Q'$, and let $C = (\mathcal{T}, (Z_e)_{e \in \mathcal{T}}, ([\tau[y_e])_{e \in \mathcal{T}}) \in FS(q)$ such that $T = (\mathcal{T}, (Z_e)_{e \in \mathcal{T}})$. Let $\mathcal{T}^*$ be the trunk of T^*. We show how to compute characteristics $C^* = (\mathcal{T}^*, (Z_e^*)_{e \in \mathcal{T}^*}, (\tau[y_e^*])_{e \in \mathcal{T}^*})$ for $FS(p)$. We consider two cases.

  1. The trunks $\mathcal{T}$ and $\mathcal{T}^*$ are the same. In this case we can proceed as described in section 13.9. For each edge of the trunk we change the typical list as described in step 4 of the algorithm given in that section (page 168).

  2. The trunks are different. In that case, the trunk $\mathcal{T}^*$ contains a leaf say a, which is not a leaf of the trunk $\mathcal{T}$. Let b be the neighbor of a in $\mathcal{T}^*$. In general, when b is of degree three in $\mathcal{T}^*$, b is not a node in $\mathcal{T}$. In that case let c and d be the other neighbors of b in $\mathcal{T}^*$. Notice that c and d are adjacent in $\mathcal{T}$. Let Z_b be the interval corresponding with node b in T^*. Let

$$Z_e = (Z_e^{(c)}, \ldots, Z_e^{(b)}, \ldots, Z_e^{(d)})$$

be the interval model for $e = (c, d)$ in T. In T^* this interval model is split in two parts

$$Z_{e_1}^* = (Z_{e_1}^{(c)}, Z_{e_1}^{(c+1)}, \ldots, Z_{e_1}^{(b)}) \quad \text{for } e_1 = (c, b)$$
$$Z_{e_2}^* = (Z_{e_2}^{(b)}, Z_{e_2}^{(b+1)}, \ldots, Z_{e_2}^{(d)}) \quad \text{for } e_2 = (b, d)$$

Consider the typical list for e in T:

$$\tau[y_e] = (\tau(y_e^{(c)}), \ldots, \tau(y_e^{(b)}), \ldots, \tau(y_e^{(d)}))$$

The typical sequence

$$\tau(y_e^{(c)}) = (\alpha_1, \ldots, \alpha_s)$$

for the interval $Z_e^{(b)}$, is split into two parts *in all possible ways* (each split gives a characteristic for $FS(p)$):

$$\tau_1 = (\alpha_1, \ldots, \alpha_f) \wedge \tau_2 = (\alpha_f, \ldots, \alpha_s)$$

Notice that, by Lemma 13.4.13, τ_1 and τ_2 are typical sequences. The typical lists for the edges (c, b) and (b, d) in C^* are

$$\tau[y_{e_1}^*] = (\tau(y_e^{(c)}), \ldots, \tau(y_e^{(b-1)}), \tau_1) \qquad \text{for the edge } (c, b)$$
$$\tau[y_{e_2}^*] = (\tau_2, \tau(y_e^{(b+1)}), \ldots, \tau(y_e^{(d)})) \qquad \text{for the edge } (b, d)$$

When the node b is a node in T, the interval model and typical list are not split.

Finally, we have to describe the typical sequence for the edge (a, b) in $f = (a, b)$. Consider the interval model $Z_f = (Z_f^{(1)}, \ldots, Z_f^{(r)})$ for this edge in T. The typical sequence $\tau(y_f^{*(i)}) = (|Z_f^{(i)}|)$ (i.e. consists of one element).

Step 4 Stop. The computation of $FS(p)$ is completed.

In Step 3 of the algorithm, we claim that the trunks T and T^* differ only in some specified way. We start by proving this.

Lemma 14.5.1 *If the trunks T and T^* are different, then there is exactly one leaf a of T^* which is not a leaf of T. Let b be the neighbor of a in T^*. If b is of degree three in T^* then b is not a node in T. In this case the two other neighbors of b are adjacent in T.*

Proof. Let Y^* be a minimal tree-decomposition for the subgraph induced by X_p with tree-model $(T^*, (Z_e)_{e \in T})$. The trunk T is obtained by removing x from all subsets and then computing the trunk of the result. Assume the trunks are not the same. Then clearly, there must be a leaf in T^* which is not a node of T. Hence the subset corresponding with a is not maximal in T. It follows that x is contained only in this subset. $\square$

Theorem 14.5.1 *Each element $C^* = (T^*, (Z_e^*)_{e \in T^*}, (\tau[y_e^*])_{e \in T^*}) \in FS(p)$ is the characteristic of a partial tree-decomposition rooted at node p.*

Proof. Let $C = (\mathcal{T}, (Z_e)_{e \in \mathcal{T}}, (\tau[y_e])_{e \in \mathcal{T}})$ be the characteristic in FS(q) from which C^* is computed by the algorithm. Since FS(q) is a full set of characteristics for q, there exists a partial tree-decomposition $Y = (SY, TY)$ rooted at q with characteristic C. By Lemma 13.7.2 we may assume that $[y_e] \in E(\tau[y_e])$ for every edge $e \in \mathcal{T}$. We show how to compute a partial tree-decomposition rooted at p with characteristic C^*. If $\mathcal{T} = \mathcal{T}^*$ the result follows from Theorem 13.9.1. Hence assume the trunks are different. Let a be the leaf of $\mathcal{T}^*$ which is not in $\mathcal{T}$ and let b be the neighbor of a. Assume b is not an element of $\mathcal{T}$. Then b has two other neighbors c and d which are adjacent in $\mathcal{T}$. Let e be the edge (c, d) in $\mathcal{T}$. The algorithm splits the characteristic sequence $\tau(y_e^{(b)})$ into two parts:

$$\tau_1 = (\alpha_1, \ldots, \alpha_f) \wedge \tau_2 = (\alpha_f, \ldots, \alpha_s)$$

Since $[y_e] \in E(\tau[y_e])$ we know that $y_e^{(c)} \in E(\tau(y_e^{(c)}))$. Hence we can split the sequence $Y_e^{(c)}$ in two parts $Y_{e_1}^*$ and $Y_{e_2}^*$ such that τ_1 is the characteristic sequence of $y_{e_1}^*$ and τ_2 is the characteristic sequence of $y_{e_2}^*$. Let b_0 be the node in TY, corresponding with the split of $Y_e^{(c)}$. We make a new tree TY* by making a new path P adjacent to b_0. Let f be the edge (a, b) in $\mathcal{T}^*$. Each node i in P corresponds to a subset $Z_f^{*(i)}$. The corresponding subset in Y* is equal to $Z_f^{*(i)}$. It is easily checked that Y* is a partial tree-decomposition rooted at p with characteristic C^*. $\qquad\square$

Theorem 14.5.2 *Let* Y *be a partial tree-decomposition rooted at* p *of width at most* ℓ. *There exists a partial tree-decomposition* Y' *such that* $Y' \prec Y$ *and such that* $C(Y') \in FS(p)$.

Proof. We may assume Y is minimal. Let Y^0 be the subdecomposition of Y rooted at q. Since FS(q) is a full set of characteristics there exists a partial tree-decomposition $Y_0 \prec Y^0$ such that $C(Y_0) \in FS(q)$. Clearly we may assume that Y_0 is minimal. Let $C(Y_0) = (\mathcal{T}, (Z_e)_{e \in \mathcal{T}}, (\tau[y_e])_{e \in \mathcal{T}})$. We may assume that $[y_e] \in E(\tau[y_e])$ for all edges $e \in \mathcal{T}$. Since $Y_0 \prec Y^0$, there exist partial tree-decompositions Y_0^* and Y^{0*} such that $Y_{0e}^* \in E(Y_{0e})$, $Y_e^{0*} \in E(Y_e^0)$ and $[y_{0e}^*] \leq [y_e^{0*}]$ for every edge $e \in \mathcal{T}$. Since Y^0 is the restriction of Y, there exists a partial tree-decomposition Y* with the same characteristic as Y such that $[y_e^*] \in E([y_e])$ and such that $[y_e^*]$ and $[y_e^{0*}]$ have the same length in the strong sense for every edge $e \in \mathcal{T}$. Assume the trunk of Y and Y^0 are different. Since Y and Y_0 are minimal there is a simple path P in the tree TY of Y which is not present in the tree TY_0 of Y_0. Let i be the node in TY_0^* which is adjacent to a node of P in TY*. Make P adjacent to i in TY_0. $\qquad\square$

Chapter 15

Recognizing treewidth-bounded graphs

In this chapter we describe, for each constant k, a linear time algorithm which, given a graph G determines whether the treewidth of G is at most k. This algorithm was found by Bodlaender [21].

15.1 Leaf-path-collections

We start with some easy observations for trees and k-trees. Recall the definition of a clique-tree-decomposition for a chordal graph, Definition 7.2.1 (on 81).

Let $Z = (V, C)$ be a k-tree with n vertices and let $D = (T, S)$ be a clique-tree-decomposition with $T = (I, F)$. Notice that $|I| = n - k$ (which is the number of $k + 1$-cliques in Z by Lemma 2.1.10).

Definition 15.1.1 *A path in* T *is a (connected) subtree such that all internal vertices have degree two in* T.

Let $\ell \geq 1$ be some integer (to be determined later).

Definition 15.1.2 *A leaf-path-collection of* T *with paths of length* ℓ *(i.e., paths with* ℓ *nodes) is a set of leaves and paths of length* ℓ, *such that no leaf lies on a path and no two paths have a node in common.*

Lemma 15.1.1 T *has a leaf-path-collection* $\mathcal{P}$ *with at least* $\frac{n-k}{2\ell}$ *elements.*

Proof. Let b be the number of leaves and δ be the number of vertices of degree at least three. Hence the number of maximal paths in T, containing only nodes of degree two, is at most $b + \delta - 1$ (some paths may be empty). In each of these maximal paths we can choose paths for the leaf-path-collection $\mathcal{P}$ such that at most $\ell - 1$ nodes of degree two are not in a path of $\mathcal{P}$. Hence there are at most $(b+\delta-1)(\ell-1)$ nodes in T of degree two that are not in a path of $\mathcal{P}$. Hence at least $n-k-(\delta+b)-(\delta+b-1)(\ell-1) \geq n-k-(\delta+b)\ell$ nodes of degree two are in a path of $\mathcal{P}$. It follows that we can choose $\mathcal{P}$ such that the number of paths in $\mathcal{P}$ is at least $\frac{n-k}{\ell} - (\delta+b)$. Hence there exists a leaf-path-collection $\mathcal{P}$ with

$$|\mathcal{P}| \geq \max(b, \frac{n-k}{\ell} - \delta) \geq \max(b, \frac{n-k}{\ell} - b)$$

since, obviously, $\delta \leq b$. The lemma follows.   □

15.2 Matchings

Definition 15.2.1 *A* matching *in a graph* $G = (V, E)$ *is a set* M *of edges such that no two edges of* M *have a vertex in common. A matching is called* maximal *if it not contained in a larger set.*

Let $d \geq k+1$ be some constant (to be determined later).

Definition 15.2.2 H *is the set of vertices of high degree, i.e., those vertices with degree larger than* d. L *is the set of low degree vertices, i.e., those with degree at most* d. L *is partitioned further into sets* L_I *and* L_M. L_M *are those vertices of* L *which have at least one neighbor in* L. L_I *are those vertices of* L *without any neighbor in* L

Remark. Notice that vertices of L_I are isolated in $G[L]$.

Lemma 15.2.1 *If* G *has treewidth at most* k, *then* $|H| \leq \frac{2nk}{d}$.

Proof. The number of edges in G is at least $|H|d/2$. By Lemma 2.1.10, the number of edges can not be larger than nk.   □

Lemma 15.2.2 *Every maximal matching in* G *has at least* $\frac{|L_M|}{2d}$ *edges.*

Proof. Let M be a maximal matching. Let U be the set of end points of edges in M that are vertices of L_M. Then clearly $|U| \leq 2|M|$. Count the number of edges between U and $L_M \setminus U$ in two different ways. Since vertices of U have degree at most d, and since at least one neighbor is not in $L_M \setminus U$, this number of edges is at most $|U|(d-1)$. On the other hand, since M is a maximal matching, $L_M \setminus U$ is an independent set. Hence every vertex of $L_M \setminus U$ is adjacent to at least one vertex of U. Hence the number of edges between U and $L_M \setminus U$ is at least $|L_M \setminus U|$. We obtain $|L_M| \leq |U|d$. Hence, $|M| \geq \frac{|U|}{2} \geq \frac{|L_M|}{2d}$.   □

15.3 Dangling vertices

In this section, let $G = (V, E)$ be a graph with treewidth at most k, and let $Z = (V, C)$ be a k-tree embedding of G. Consider a clique-tree-decomposition $D = (T, S)$ of Z with $T = (I, F)$ and $S = \{X_i \mid i \in I\}$. Hence D is a tree-decomposition for G.

We will denote the neighbors of a vertex $x \in V$ in G by $N_G(x)$ and its neighbors in Z by $N_Z(x)$.

Let $\mathcal{P}$ be a leaf-path-collection of T with paths of length ℓ, with at least $\frac{n-k}{\ell}$ elements. For a path $P \in \mathcal{P}$ let $I(P)$ be the set of nodes. For a path P, let $\overset{\circ}{P}$ denote the interior, (with nodes $I(\overset{\circ}{P})$). For a subset $J \subseteq I$ of nodes let $X(J) = \cup_{i \in J} X_i$.

Definition 15.3.1 *A vertex* $x \in L_I$ *is called* dangling *in* D *if there is a node* $i \in I$ *such that* $N_G(x) \subseteq X_i$.

Lemma 15.3.1 *If* $d \geq k + \ell$ *and* $\ell \geq k^2 + 3k + 4$ *then every path* $P \in \mathcal{P}$ *has at least one node* $i \in I(\overset{\circ}{P})$ *such that* X_i *has a vertex which is either dangling or an element of* L_M *and which is not in* X_j *for any* $j \notin I(\overset{\circ}{P})$.

Proof. Let $P = (1, 2, \ldots, \ell)$ be a path in $\mathcal{P}$. Notice that $|X(I(P))| = k + \ell \leq d$, hence, all vertices of $X(I(P)) \setminus (X_1 \cup X_\ell)$ are of low degree. Let Q be this set of vertices. If one of these vertices is in L_M then we are done since they do not occur in X_1 or X_ℓ. Hence, all vertices of Q are only adjacent to vertices in $X_1 \cup X_\ell$.

Let $H \cap X_1 = \{v_1, \ldots, v_r\}$. Then clearly, $r \leq k + 1$. Assume v_i occurs in subsets $X_1, \ldots, X_{w_i}$, and let $w_1 \leq \ldots \leq w_r$.

Case 1 Assume some vertex of Q occurs only in *one* subset X_j, for some $2 \leq j \leq \ell - 1$. In that case it is dangling by definition.

Case 2 Assume some vertex $x \in Q$ occurs only in some subsets of $X_{w_i+1}, \ldots, X_{w_{i+1}}$ for some i. In that case all neighbors of x are contained in $X_{w_{i+1}}$. Hence it is dangling.

Case 3 If one the former cases would not occur, all vertices of $X(I(P))$ had to be in at least one subset of $X_1, X_{w_1}, \ldots, X_{w_r}, X_\ell$. But the number of vertices in these last subsets can be at most $(k+1)(k+3) = k^2 + 4k + 3 < k + \ell = |X(I(P))|$, which is a contradiction.

$\square$

Theorem 15.3.1 *If* $d \geq k + \ell$ *and* $\ell \geq k^2 + 3k + 4$ *then there are at least* $\frac{n-k}{2\ell}$ *vertices that are in* L_M *or dangling.*

Proof. Consider a leaf-path-collection $\mathcal{P}$ with at least $\frac{n-k}{2\ell}$ elements (Lemma 15.1.1). If i is a leaf of T, then X_i contains a vertex x that is simplicial in Z. Hence $N_G(x) \subseteq N_Z(x) \subset X_i$. Hence this vertex is in L_M or dangling in D.

By Lemma 15.3.1 every path $P \in \mathcal{P}$ has a node $i \in I(\overset{\circ}{P})$ such that X_i has a vertex which is in L_M or dangling in D. □

15.4   Partly-filled graphs

Definition 15.4.1 *The* partly-filled graph G' *is the graph obtained from* G *as follows. If two non adjacent vertices* x *and* y *of* H *have at least* $k+1$ *common neighbors in* L, *then* x *and* y *are adjacent in* G'.

Lemma 15.4.1 *If the treewidth of* G *is at most* k, *then* G' *is a subgraph of every triangulation of* G *into a* k-tree.

Proof. Notice that, if (A, B) is a complete bipartite subgraph of G, then in any triangulation either A or B is a clique, otherwise there would be a chordless 4-cycle (see Lemma 2.2.1 on page 18). Now if x and y are two vertices of H then $\{x, y\}$ and the set of common neighbors form a complete bipartite subgraph. Hence either $\{x, y\}$ is an edge in a triangulation, or the set of common neighbors is a clique. If the set of common neighbors has at least $k+1$ elements, then this cannot be a clique in a k-tree embedding, otherwise there would be a $k+2$-clique. □

In the rest of this section we obtain an upper bound on the number of vertices that are not simplicial in the partly-filled graph G'. In fact, we look only for simplicials in L_I.

Again let $Z = (V, C)$ be a k-tree embedding of G (hence also of G'). Consider again a clique-tree-decomposition $D = (T, S)$ of Z with $T = (I, F)$ and $S = \{X_i \mid i \in I\}$.

Definition 15.4.2 *A subset* $Y \subseteq Z$ *is a* clique-set *in* D *if there is a node* $i \in I$ *such that* $Y \subseteq X_i$. *A* clique-set *is* maximal *if it is not properly contained in another clique-set.*

Lemma 15.4.2 *There are at most* $|H|$ *maximal clique-sets.*

Proof. Consider the induced subgraph $Z[H]$. Each maximal clique-set is a maximal clique in $Z[H]$. The number of maximal cliques in $Z[H]$ is bounded by the number of vertices since this graph is chordal [66]. □

Theorem 15.4.1 *There are at most $\frac{1}{2}|H|k^2(k+1)$ dangling vertices not simplicial in the partly-filled graph G'.*

Proof. Consider the dangling vertices in L_I that are not simplicial in G'. For each, choose a pair of neighbors in H that are not adjacent. Notice that such a pair of neighbors is contained in some maximal clique-set.

Each non-adjacent pair of vertices can be assigned to at most k dangling vertices, by definition of the partly-filled graph. By lemma 15.4.2 there are at most $|H|$ maximal clique-sets, and each has at most $\binom{k+1}{2}$ non-adjacent pairs of vertices. Hence there are at most $|H|k\binom{k+1}{2}$ dangling vertices not simplicial in G'. $\qquad\square$

15.5 A linear time algorithm finding a tree-decomposition

In this section we describe the linear time algorithm for finding an optimal tree-decomposition for a graph with bounded treewidth. We first state our main theorem.

Theorem 15.5.1 *There exist constants $\epsilon = \epsilon(k) > 0$ and $\delta = \delta(k) > 0$ such that for every graph G with at least $k+1$ vertices at least one of the following properties holds.*

1. *The treewidth of G is larger than k.*

2. *Every maximal matching has at least ϵn edges.*

3. *There are at least δn vertices in L_I that are simplicial in the partly-filled graph G'.*

Proof. Choose $\ell \geq k^2 + 3k + 4$ and $d \geq k + \ell$. By Theorem 15.3.1 there are at least $\frac{n-k}{2\ell} \geq \frac{n}{2\ell(k+1)}$ vertices in L_M or dangling.

By Theorem 15.4.1 and Lemma 15.2.1, among the dangling vertices there are at most $\frac{nk^3(k+1)}{d}$ vertices that are not simplicial in the partly-filled graph. Hence we find at least

$$n\left(\frac{1}{2\ell(k+1)} - \frac{|L_M|}{n} - \frac{k^3(k+1)}{d}\right)$$

vertices that are simplicial in the partly-filled graph G'.

Choose $d > 2\ell k^3(k+1)^2$,

$$0 < \epsilon < \frac{1}{2d}\left(\frac{1}{2\ell(k+1)} - \frac{k^3(k+1)}{d}\right)$$

$$\text{and } 0 < \delta < \frac{1}{2\ell(k+1)} - 2\epsilon d - \frac{k^3(k+1)}{d}$$

Then, by Lemma 15.2.2 if $|L_M| \geq 2\epsilon dn$ every maximal matching has at
least ϵn edges. On the other hand, if $|L_M| < 2\epsilon dn$ we find at least δn
simplicial vertices in the partly-filled graph.   □

The following two basic observations lead to the algorithm. Consider
a maximal matching $M = \{(x_1, y_1), \ldots, (x_t, y_t)\}$. Let $G(M)$ be the graph
obtained from G by contracting the edges of M (see page 3). Hence, if
G has n vertices, $G(M)$ has $n - t$ vertices. Since $G(M)$ is a minor of
G, the treewidth of $G(M)$ is at most equal to the treewidth of G. Now,
consider a tree-decomposition D' for $G(M)$ with width at most k. We can
turn this into a tree-decomposition D for G as follows. For each subset X
in the tree-decomposition D' containing a contracted edge (x_i, y_i), replace
this contracted edges by its two end vertices in D. Obviously, D is a tree-
decomposition for G with width at most $2k + 1$. Since k is a constant, we
can use D as an approximate tree-decompostion and apply the algorithm of
chapter 14 to obtain a tree-decomposition for G with width k.

Let S be a set of simplicial vertices of the partly-filled graph G'. Consider
the graph $G(S) = G'[V \setminus S]$. Clearly, if some vertex of S has more than k
neighbors, the treewidth of G is larger than k, since G' has clique number
larger than k. If $G(S)$ has a k-tree embedding H^*, we can obtain a k-tree
embedding H for G, by adding the vertices of S.

A more precise description of the algorithm is the following. Let k, d,
ℓ, ϵ and δ be constants as mentioned in Theorem 15.5.1

- **algorithm:** determines whether the treewidth of G is at most k and if
  so computes a tree-decompostion D with width at most k.

- If $n \leq k + 1$, output, that the treewidth is at most k; **stop**.

- Check if the number of edges is at most $nk - \frac{1}{2}k(k + 1)$. If the
  number of edges is larger, then the treewidth is larger than k; **stop**.
  (Lemma 2.1.12 page 13).

- Determine a maximal matching M.

- If $|M| \geq \epsilon n$ then do the following:

  - Contract all edges of M; obtaining the graph $G(M)$.

  - Recursively apply the algorithm to $G(M)$.

  - If $G(M)$ has treewidth larger than k then **stop**; G also has
    treewidth larger than k.

  - Otherwise, the algorithm computes a tree-decomposition D'
    for $G(M)$ with width at most k. Change this into a tree-
    decomposition D for G with width at most $2k + 1$.

- Use D and the algorithm of Chapter 14 to either decide that the treewidth of G exceeds k or to determine a tree-decomposition of G with width at most k.

- Assume $|M| < \epsilon n$. Then compute the partly-filled graph G'.

- Compute the set S of all vertices in L_I that are simplicial in G'. If some vertex x of S has more than k neighbors, then **stop**; the treewidth of G is larger than k.

- If $|S| < \delta n$, then the treewidth of G is larger than k; **stop**.

- Otherwise, compute $G(S) = G'[V \setminus S]$.

- Recursively apply the algorithm to $G(S)$. If the treewidth of $G(S)$ is larger than k, then so is the treewidth of G: **stop**.

- Otherwise, the algorithm returns a tree-decomposition D' of $G(S)$ with width at most k. Use D' to obtain a tree-decomposition D for G with width at most k.

Notice that the algorithm is recursively applied to $G(M)$ which has at most $(1 - \epsilon)n$ vertices or to graph $G(S)$ which has at most $(1 - \delta)n$ vertices. Since both ϵ and δ are positive, the algorithm needs only linear time as long as all non recursive steps are implemented in linear time.

We discuss some details.

Clearly, a maximal matching M can be obtained in linear time. Let $M = \{(x_1, y_1), \ldots, (x_t, y_t)\}$. The graph $G(M)$ can be obtained as follows. First go through all adjacency lists and replace all occurrences of each y_i by x_i. Remove all vertices y_i $(i = 1, \ldots, t)$ from the graph. The next step is to remove duplicate edges. This can easily be done in linear time by sorting all adjacency lists.

Next we describe how the partly-filled graph G' can be computed. the same method was also used in [21]. First, determine the sets H, L_I and L_M. For every pair of vertices in H with at least $k+1$ common neighbors in L, an edge has to be added (if not yet present). Let the vertices of H be numbered $v_1, \ldots, v_t$.

Make a queue Q containing entries of the form $((v_i, v_j), -)$ or $((v_i, v_j), x)$, where $i < j$, v_i and v_j are vertices of H and x is a common neighbor of v_i and v_j in L. For every edge (v_i, v_j) with $i < j$ in H, put an entry $((v_i, v_j), -)$ in Q.

Next, for every vertex $x \in L$, and for every pair of vertices (v_i, v_j) with $i < j$ in $N_G(x) \cap H$, put an entry $((v_i, v_j), x)$ in the queue Q.

Use bucket sort twice, to sort the entries of Q in alphabetically increasing order of the pairs (v_i, v_j). Hence, having done this, all entries of the form $((v_i, v_j), \ldots)$ occur consecutively. Now it is easy to count for each pair (v_i, v_j)

the number of common neighbors in L. If this number is more than $k+1$, and if the entry $((v_i, v_j), -)$ is not present in Q, we add the edge (v_i, v_j) in the graph. In this way we can obtain the partly-filled graph G' in linear time.

Finally, we have to determine the set S of vertices in L_I that are simplicial in G'. For each vertex $x \in L_I$ we count the number of edges in $N_{G'}(x)$ as follows. Scan the sorted queue Q. For each edge (v_i, v_j) and for each $x \in L_I$ such that $((v_i, v_j), x) \in Q$ and such that either $((v_i, v_j), -) \in Q$ or v_i and v_j have at least $k+1$ common neighbors in L, add one to the counter of x. In this way we can determine S in linear time. Removing all vertices of S from G' gives the graph $G(S)$.

Bibliography

[1] Alon, N., P. Seymour and R. Thomas, A separator theorem for graphs with an excluded minor and its applications, *Proceedings of the 22nd ACM Symposium on Theory of Computing* (1990), pp. 293–299.

[2] Anand, R., H. Balakrishnan and C. Pandu Rangan, Treewidth of distance hereditary graphs. Manuscript.

[3] Arnborg, S., Efficient algorithms for combinatorial problems on graphs with bounded decomposability — A survey, *BIT* **25**, (1985), pp. 2–23.

[4] Arnborg, S., D. G. Corneil and A. Proskurowski, Complexity of finding embeddings in a k-tree, *SIAM J. Alg. Disc. Meth.* **8**, (1987), pp. 277–284.

[5] Arnborg, S., J. Lagergren and D. Seese, Easy problems for tree-decomposable graphs, *J. Algorithms* **12**, (1991), pp. 308–340.

[6] Arnborg, S. and A. Proskurowski, Characterization and recognition of partial 3-trees, *SIAM J. Alg. Disc. Meth.* **7**, (1986), pp. 305–314.

[7] Arnborg, S. and A. Proskurowski, Linear time algorithms for NP-hard problems restricted to partial k-trees, *Disc. Appl. Math.* **23**, (1989), pp. 11–24.

[8] Bandelt, H. J. and H. M. Mulder, Distance-Hereditary Graphs, *Journal of Combinatorial Theory*, Series **B 41**, (1986), pp. 182–208.

[9] Beineke, L. W., and R. E. Pippert, The enumeration of labelled 2-trees, *Notices Amer. Math. Soc.* **15**, (1968), pp. 384.

[10] Beineke, L. W. and R. E. Pippert, The number of labeled k-dimensional trees, *J. Combinatorial Theory* **6**, (1969), pp. 200–205.

[11] Benzaken, C., Y. Crama, P. Duchet, P. L. Hammer and F. Maffray, More characterizations of triangulated graphs, *J. of Graph Theory*, **14**, (1990), pp. 413–422.

[12] Berge, C., Farbung von Graphen deren sämtliche bzw. deren ungerade Kreise starr sind, *Wiss. Z. Martin-Luther-Univ., Halle-Wittenberg Math.-Natur, Reihe*, pp. 114-115.

[13] Berge, C., Les problèmes de coloration en théorie des graphes, *Publ. Inst. Statist. Univ. Paris* **9**, (1960), pp. 123–160.

[14] Berge, C., Sur une conjecture relative aux problème des codes optimaux, *Comm.* 13*ieme Assemblee Gen. URSI*, Tokyo, (1962).

[15] Berge, C. and C. Chvatal, *Topics on Perfect Graphs*, Ann. Disc. Math. **21**, 1984.

[16] Bierstone, E., Cliques and generalized cliques in a finite linear graph, Unpublished report, University of Toronto, 1967.

[17] Bodlaender, H., Classes of graphs with bounded treewidth, Technical Report RUU-CS-86-22, Department of Computer Science, Utrecht University, 1986.

[18] Bodlaender, H. L., Dynamic programming algorithms on graphs with bounded treewidth, *Proceedings of the 15th International Colloquium on Automata, Languages and Programming*, Springer-Verlag, Lecture Notes in Computer Science 317, (1988), pp. 105–119.

[19] Bodlaender, H. L., Achromatic number is NP-complete for cographs and interval graphs, *Information Processing Letter* **31**, (1989), pp. 135–138.

[20] Bodlaender, H. L., A tourist guide through treewidth, Technical report RUU-CS-92-12, Department of Computer Science, Utrecht University, Utrecht, The Netherlands, (1992).

[21] Bodlaender, H., A linear time algorithm for finding tree-decompositions of small treewidth, *Proceedings of the 25th Annual ACM Symposium on Theory of Computing*, (1993), pp. 226–234.

[22] Bodlaender, H., J. S. Deogun, K. Jansen, T. Kloks, D. Kratsch, H. Müller and Z. Tuza, Rankings of graphs. To appear in *Proceedings of the International Workshop on Graph-Theoretic Concepts in Computer Science WG'94*.

[23] Bodlaender, H., M. Fellows and T. Warnow, Two strikes agains Perfect Phylogeny, *Proceedings of the 19th International colloquium on Automata, Languages and Programming*, Springer-Verlag, Lecture Notes in Computer Science 623, (1992), pp. 273–283.

[24] Bodlaender, H., L. van der Gaag and T. Kloks, Some remarks on minimum edge and minimum clique triangulations, Unpublished result.

[25] Bodlaender, H., J. Gilbert, H. Hafsteinsson and T. Kloks, Approximating treewidth, pathwidth and minimum elimination tree height, *Proceedings* 17th *International Workshop on Graph-Theoretic Concepts in Computer Science WG'91*, Springer-Verlag, Lecture Notes in Computer Science 570, (1992), pp. 1–12. To appear in *J. Algorithms*.

[26] Bodlaender, H. and T. Kloks, Fast Algorithms for the TRON game on trees, Technical Report RUU-CS-90-11, Department of Computer Science, Utrecht University, Utrecht, The Netherlands, (1990).

[27] Bodlaender, H. and T. Kloks, Better algorithms for the pathwidth and treewidth of graphs, *Proceedings of the* 18th *International colloquium on Automata, Languages and Programming*, Springer-Verlag, Lecture Notes in Computer Science 510, (1991), pp. 544–555.

[28] Bodlaender, H. and T. Kloks, A simple linear time algorithm for triangulating three-colored graph, 9th *Annual Symposium on Theoretical Aspects of Computer Science*, Springer-Verlag, Lecture Notes in Computer Science 577, (1992), pp. 415–423. To appear in *J. Algorithms*.

[29] Bodlaender, H. and T. Kloks, Efficient and constructive algorithms for the pathwidth and treewidth of graphs, Technical Report RUU-CS-93-27, Department of Computer Science, Utrecht University, Utrecht, The Netherlands, (1993).

[30] Bodlaender, H., T. Kloks and D. Kratsch, *Proceedings of the* 20th *International colloquium on Automata, Languages and Programming* Springer Verlag, Lecture Notes in Computer Science 700, (1993), pp. 114–125.

[31] Bodlaender, H., T. Kloks, D. Kratsch and H. Müller, Treewidth and pathwidth of cotriangulated graphs, unpublished.

[32] Bodlaender, H. and R. H. Möhring, The pathwidth and treewidth of cographs, *Proceedings* 2nd *Scandinavian Workshop on Algorithm Theory*, Springer-Verlag, Lecture Notes in Computer Science 447, (1990), pp. 301–309.

[33] Bollobás, B., *Random Graphs*, Academic Press, New York, 1985.

[34] Bollobás, B., P. A. Catlin and P. Erdős, Hadwiger's conjecture is true for almost every graph, *Europ. J. Combinatorics* **1**, (1980), pp. 195–199.

[35] Booth, K. and G. Lueker, Testing for the consecutive ones property, interval graphs, and graph planarity testing using PQ-tree algorithms, *J. of Computer and System Sciences* **13**, (1976), pp. 335–379.

[36] Bouchet, A., A polynomial algorithm for recognizing circle graphs, *C. R. Acad. Sci. Paris, Sér. I Math.* **300**, (1985), pp. 569–572.

[37] Bouchet, A., Reducing prime graphs and recognizing circle graphs, *Combinatorica* **7**, (1987), pp. 243–254.

[38] Brandstädt, A., Special graph classes — a survey, Schriftenreihe des Fachbereichs Mathematik, SM-DU-199 (1991), Universität Duisburg Gesamthochschule.

[39] Brandstädt, A. and D. Kratsch, On the restriction of some NP-complete graph problems to permutation graphs, *Fundamentals of Computation Theory, proc. FCT*, Springer-Verlag, Lecture Notes in Computer Science 199, (1985), pp. 53–62.

[40] Brandstädt, A. and D. Kratsch, On domination problems for permutation and other perfect graphs, *Theor. Comput. Sci.* **54**, (1987), pp. 181–198.

[41] Bron, S. and J. Kerbosch, Algorithm 457 — Finding all cliques of an undirected graph, *Comm. of ACM* **16** (1973), pp. 575.

[42] Burnside, W., *Theory of groups of finite order* (second edition), Cambridge University Press, Cambridge, (1911).

[43] Chang, M.-S. and F.-H. Wang, Efficient algorithms for the maximum weight clique and maximum weight independent set problems on permutation graphs, *Information Processing Letters* **43**, (1992), pp. 293–295.

[44] Christofides, N., An Algorithm for the chromatic number of a graph, *Computer J.* **14** (1971), pp. 38–39.

[45] Christofides, N., *Graph Theory, An Algorithmic Approach*, Academic Press, New York, 1975.

[46] Cohen, J. E., J. Komlós and T. Mueller, The probability of an interval graph, and why it matters, *Proc. of Symposia in Pure Math.* **34**, (1979), pp. 97–115.

[47] Corneil, D. G., Y. Perl, L. Stewart, Cographs: recognition, applications and algorithms, *Congressus Numerantium* **43**, (1984), pp. 249–258.

[48] Courcelle, B., The monadic second-order logic of graphs I: Recognizable sets of finite graphs, *Information and Computation* **85**, (1990), pp. 12–75.

[49] Courcelle, B., The monadic second-order logic of graphs III: Treewidth, forbidden minors and complexity issues, Report 8852, University Bordeaux **1**, (1988). To appear in *Informatique Théoretique et Applications*.

[50] Courcelle, B., Graph rewriting: an algebraic and logical approach. In J. van Leeuwen, editor, *Handbook of Theoretical Computer Science, Vol. B*, Amsterdam, Elsevier Science Publ. (1990), pp. 192–242.

[51] Courcelle, B., Proceedings of the International Workshop on Graphs and Graph Transformations. Part 1: Tree-structured graphs, Forbidden configurations and graph algorithms. March (1991).

[52] Dahlhaus, E., Chordale Graphen im besonderen Hinblick auf parallele Algorithmen, Habilitationsschrift, Bonn, (1989).

[53] Damaschke, P. and H. Müller, Hamiltonian circuits in convex and chordal bipartite graphs, submitted to Discrete Mathematics.

[54] Deo, N., M. S. Krishnamoorty, and M. A. Langston, Exact and approximate solutions for the gate matrix layout problem, *IEEE Transactions on Computer-Aided Design* **6**, (1987), pp. 79–84.

[55] Deogun, J. S., T. Kloks, D. Kratsch and H. Müller, On vertex ranking for permutation and other graphs, 11th *Annual Symposium on Theoretical Aspects of Computer Science*, Springer-Verlag, Lecture Notes in Computer Science 775, (1994), pp. 747–758.

[56] Dirac, G. A., On rigid circuit graphs, *Abh. Math. Sem. Univ. Hamburg* **25**, (1961), pp. 71–76.

[57] Ellis, J. A., I. H. Sudborough and J. Turner, Graph separation and search number, Technical report DCS-66-IR, University of Victoria, (1987).

[58] Erdös, P., A. Gyárfás, E. T. Ordman and Y. Zalcstein, The size of chordal, interval and threshold subgraphs, *Combinatorica* **9**, (1989), pp. 245–253.

[59] Even, S., A. Pnueli and A. Lempel, Permutation graphs and transitive graphs, *J. Assoc. Comput. Mach.* **19**, (1972), pp. 400–410.

[60] Farber, M., Characterizations of strongly chordal graphs, *Disc. Math.* **43**, (1983), pp. 173–189.

[61] Farber, M. and M. Keil, Domination in permutation graphs, *J. Algorithms* **6**, (1985), pp. 309–321.

[62] Fishburn, P. C., An interval graph is not a comparability graph, *J. Combin. Theory* **8**, (1970), pp. 442–443.

[63] Foata, D., Enumerating k-trees, *Discrete Mathematics* **1**, (1971), pp. 181–186.

[64] Foster, R. M., The number of series-parallel networks, *Proc. Int. Congress Math. 1950* **1**, (1952) pp. 642.

[65] Foster, R. M., Topologic and algebraic considerations in network synthesis, *Proceedings of the symposium on Modern Network Synthesis*, (1952), pp. 8–18.

[66] Fulkerson, D. R. and O. A. Gross, Incidence matrices and interval graphs, *Pacific J. Math.* **15**, (1965), pp. 835–855.

[67] Gabor, C. P., W. L. Hsu and K. J. Supowit, Recognizing circle graphs in polynomial time, *26th Annual IEEE Symposium on Foundations of Computer Science*, (1985).

[68] Gallai, T., Transitive orientierbaren Graphen, *Acta Math. Sci. Hung.* **18**, (1967), pp. 25–66.

[69] Garbe, R., Tree-width and path-width of comparability graphs of interval orders. To appear in *Proceedings of the International Workshop on Graph-Theoretics Concepts in Computer Science WG'94*.

[70] Garey, M. R. and D. S. Johnson, *Computers and Intractability: A Guide to the Theory of NP-completeness*, San Francisco, 1979.

[71] Gavril, F., Algorithms on clique separable graphs, *Discrete Math.* **19**, (1977), pp. 159–165.

[72] Gilmore, P. C. and A. J. Hoffman, A characterization of comparability and interval graphs, *Canad. J. Math.* **16**, (1964), pp. 539–548.

[73] Gimbel, J., D. Kratsch and L. Stewart, On cocolourings and cochromatic numbers of graphs. To appear in *Disc. Appl. Math.*

[74] Goh, L. and D. Rotem, Recognition of perfect elimination bipartite graphs, *Information Processing Letters*, **15**, (1982), pp. 179–182.

[75] Goldberg, L. A., *Efficient algorithms for listing combinatorial structures*, Cambridge University press, 1993.

[76] Golumbic, M. C., *Algorithmic Graph Theory and Perfect Graphs*, Academic Press, New York, 1980.

[77] Golumbic, M. C., D. Rotem and J. Urrutia, Comparability graphs and intersection graphs, *Disc. Math.* **43**, (1983), pp. 37–46.

[78] Gustedt, J., On the pathwidth of chordal graphs. Technical report 221/1989, Technical University Berlin, Berlin, Germany, (1989). To appear in *Discr. Appl. Math.*

[79] Habib, M. and R. H. Möhring, Treewidth of cocomparability graphs and a new order-theoretic parameter, Technical report 336/1992, Technische Universität Berlin, Berlin, Germany, September (1992).

[80] Hajös, G., Über eine Art von Graphen, *Intern. Math. Nachr.* **11**, Problem 65.

[81] Harary, F., The number of linear, directed, rooted and connected graphs, *Trans. Amer. Math. Soc.* **78**, (1955), pp. 445–463.

[82] Harary, F., *Graph Theory*, Addison-Wesley Publ. Comp., Reading, Massachusetts, (1969).

[83] Harary, F. and E. M. Palmer, On acyclic simplicial complexes, *Mathematika* **15** (1968), pp. 115–122.

[84] Harary, F., E. M. Palmer and R. C. Read, On the cell-growth problem for arbitrary polygons, *Disc. Math.* **11** (1975), pp. 371–389.

[85] Harary, F. and R. C. Read, The enumeration of tree-like polyhexes, *Proc. Edinburgh Math. Soc.* **17** (1970), pp. 1–14.

[86] Hering, F., R. C. Read and G. C. Shephard, The enumeration of stack polytopes and simplicial clusters, *Disc. Math.* **40** (1982), pp. 203–217.

[87] Hsu, W.-L., The perfect graph conjecture on special graphs—A survey, in: *Topics on perfect graphs*, Ann. Disc. Math. **21**, (1984), pp. 103–113.

[88] Hsu, W.-L., A simple test for the consecutive ones property, *Third International Symposium, ISAAC'92*, Springer-Verlag, Lecture Notes in Computer Science 650, (1992), pp. 459–468.

[89] Johnson, D. S., The NP-completeness column: An ongoing guide, *J. Algorithms* **6**, (1985), pp. 434–451.

[90] Jordan, C., Sur les assemblages de lignes, *Journal Reine Angew. Math.*
70, (1869), pp. 185–190.

[91] Kanevsky, A., On the number of minimum size separating vertex sets in
a graph and how to find all of them, *Proceedings of the First Annual
ACM-SIAM Symposium on Discrete Algorithms*, (1990), pp. 411–
421.

[92] Kannan, S. and T. Warnow, Inferring Evolutionary History from DNA
Sequences, *Proceedings of the 31th Annual Symposium on Foundations of Computer Science*, (1990), pp. 362–371.

[93] Kannan, S. and T. Warnow, Triangulating three-colored graphs, *SIAM
J. Discr. Meth.* **5**, (1992), pp. 249–258.

[94] Kashiwabara, T. and T. Fujisawa, NP-completeness of the problem of
finding a minimum-clique-number interval graph containing a given
graph as a subgraph, *IEEE Symp. of Circuits ans Systems*, IEEE,
(1979), pp. 657–660.

[95] Klein, P., A. Agrawal, R. Ravi, and S. Rao, Approximation through multicommodity flow, *Proceedings of the 31st Annual IEEE Symposium
on Foundations of Computer Science*, (1990), pp. 726–737.

[96] Kloks, T., Enumeration of biconnected partial 2-trees. Manuscript.

[97] Kloks, T., Minimum fill-in for chordal bipartite graphs, Technical Report RUU-CS-93-11, Department of Computer Science, Utrecht University, Utrecht, The netherlands, (1993).

[98] Kloks, T., Treewidth of circle graphs, 4th *International Symposium,
ISAAC'93*, Springer-Verlag, Lecture Notes in Computer Science 762,
(1993), pp. 108–117.

[99] Kloks, T., *Treewidth*, PhD Thesis, Utrecht University, The Netherlands,
(1993).

[100] Kloks, T., Separators in graphs, Abstract in proceedings of the Ober-
wolfach meeting *Algorithmische Methoden der Diskreten Mathematik*, November 1993.

[101] Kloks, T., $K_{1,3}$-free and W_4-free graphs. To appear.

[102] Kloks, T. and H. Bodlaender, testing superperfection of k-trees, *Third
Scandinavian Workshop on Algorithm Theory, SWAT'92* Springer-
Verlag, Lecture Notes in Computer Science 621, (1992), pp. 292–303.

[103] Kloks, T. and H. Bodlaender, On the treewidth and pathwidth of permutation graphs, Technical Report RUU-CS-92-13, Department of Computer Science, Utrecht University, Utrecht, The Netherlands, (1992).

[104] Kloks, T. and H. Bodlaender, Approximating treewidth and pathwidth of some classes of perfect graphs, *Third International Symposium, ISAAC'92*, Springer-Verlag, Lecture Notes in Computer Science 650, (1992), pp. 116–125.

[105] Kloks, T. and H. Bodlaender, Only few graphs have bounded treewidth, Technical Report RUU-CS-92-35, Department of Computer Science, Utrecht University, Utrecht, The Netherlands, (1992).

[106] Kloks, T. and D. Kratsch, Treewidth of chordal bipartite graphs, 10th *Annual Symposium on Theoretical Aspects of Computer Science*, Springer-Verlag, Lecture Notes in Computer Science 665, (1993), pp. 80–89.

[107] Kloks, T. and D. Kratsch, Finding all minimal separators of a graph, 11th *Annual Symposium on Theoretical Aspects of Computer Science*, Springer-Verlag, Lecture Notes in Computer Science 775, (1994), pp. 759–768.

[108] Kloks, T., D. Kratsch and H. Müller, Dominoes, *Computing Science Notes* **94/12**, Eindhoven University of Technology, Eindhoven, The Netherlands, (1994). To appear in *Proceedings of the International Workshop on Graph-Theoretics Concepts in Computer Science WG'94*.

[109] Kloks, T., D. Kratsch and J. Spinrad, Treewidth and pathwidth of cocomparability graphs of bounded dimension, *Computing Science Notes* **93/46**, Eindhoven University of Technology, Eindhoven, The Netherlands, (1994).

[110] Kloks, T., J. van Leeuwen and R. B. Tan, An on-line algorithm for the weighted maximum increasing subsequence problem. Manuscript.

[111] König, D., *Theorie der Graphen*, Reprinted by Chelsea Publishing Company, New York, 1950.

[112] Kostochka, A. V., A lower bound for the Hadwiger number of a graph as a function of the average degree of its vertices, *Discret. Analyz, Novosibirsk* **38** (1982), pp. 37–58.

[113] Kratsch, D., T. Kloks and H. Müller, Computing the toughness and the scattering number for interval graphs and other graphs, *Publication Interne* **806**, IRISA, Rennes Cedex, France, (1994).

[114] Kratsch, D. and L. Stewart, Domination on cocomparability graphs. To appear.

[115] Lagergren, J., Efficient parallel algorithms for tree-decomposition and related problems, *Proceedings of the 31st Annual IEEE Symposium on Foundations of Computer Science*, (1990), pp. 173–182.

[116] Lagergren, J. and S. Arnborg, Finding minimal forbidden minors using a finite congruence, *Proceedings of the 18th International colloquium on Automata, Languages and Programming*, Springer-Verlag, Lecture Notes in Computer Science 510, (1991), pp. 532–543.

[117] Leeuwen, J. van, Graph algorithms. In: J. van Leeuwen (ed.) *Handbook of Theoretical Computer Science, A: Algorithms and Complexity*, Elsevier Science Publ., Amsterdam, 1990, pp. 527–631.

[118] Leighton, F. T., F. Makedon, and S. Tragoudas, Unpublished result, (1990).

[119] Leighton, T. and S. Rao, An approximate max-flow min-cut theorem for uniform multicommodity flow problems with applications to approximation algorithms, *Proceedings of the 29th Annual IEEE Symposium on Foundations of Computer Science*, (1988), pp. 422–431.

[120] Lekkerkerker, C. G. and J. Ch. Boland, Representation of a finite graph by a set of intervals on the real line, *Fund. Math.* **51**, (1962), pp. 45–64.

[121] Lovász, L., Normal hypergraphs and the perfect graph conjecture, *Discrete Math.* **2**, (1972), pp. 253–267.

[122] Lubiw, A., Doubly lexical orderings of matrices, *SIAM J. Comput.* **16**, (1987), pp. 854–879.

[123] Macmahon, P. A., The combinations of resistances, *Electrician* **28**, (1892), pp. 601-602.

[124] Matoušek, J. and R. Thomas, Algorithms finding tree-decompositions of graphs, *Journal of Algorithms* **12**, (1991), pp. 1–22.

[125] Jean, M., An interval graph is a comparability graph, *J. Combin. Theory* **7**, (1969), pp. 189–190.

[126] Minty, G. J., A simple algorithm for listing all the trees of a graph, *IEEE Trans. on Circuit Theory* **12** (1965) pp 120.

[127] Möhring, R., private communication (1992).

[128] Moon, J. W., The number of labeled k-trees, *J. Combinatorial Theory* **6** (1969), pp. 196–199.

[129] McMorris, C. F., T. Warnow, and T. Wimer, Triangulating colored graphs, *Proceedings SODA '92*. To appear in *SIAM J. Disc. Math.*. To appear in *proceedings SODA '93*.

[130] Müller, H. and A. Brandstädt, The NP-completeness of Steiner tree and dominating set for chordal bipartite graphs, *Theoretical Computer Science* **53**, (1987), pp. 257–265.

[131] Naji, W., Reconnaissance des graphes de cordes, *Discrete Mathematics* **54**, (1985), pp. 329–337.

[132] Nakano, S. and T. Nishizeki, A linear-time algorithm for c-triangulating three-colored graphs, *Trans. Institute of Elektronics, Information and Communication*, A, 377-A, (1994), pp. 543–546.

[133] Otter, R., The number of trees, *Annals of Mathematics* **3**, (1948), pp. 583–599.

[134] Palmer, E. M., On the number of labeled 2-trees, *J. Combinatorial Theory* **6**, (1969), pp. 206–207.

[135] Pnueli, A., A. Lempel, and S. Even, Transitive orientation of graphs and identification of permutation graphs, *Canad. j. Math.* **23**, (1971), pp. 160–175.

[136] Pólya, G., Kombinatorische Anzahlbestimmungen für Gruppen, Graphen und Chemische Verbindungen, *Acta Math.* **68** (1937), pp. 145–253.

[137] Pólya, G. and R. C. Read, *Combinatorial enumeration of groups, graphs, and chemical compounds*, Springer-Verlag, New York, 1987.

[138] Read, R. C. and R. E. Tarjan, Bounds on backtrack algorithms for listing cycles, paths, and spanning trees, *Networks* **5** (1975), pp. 237–252.

[139] Reed, B., Finding approximate separators and computing treewidth quickly, *Proceedings 24th Annual ACM Symposium on Theory of Computing*, (1992), pp. 221–228.

[140] Rényi, A. and C. Rényi, The Prüfercode for k-trees, *Proc. Colloq. on Combinatorial structures and their applications*, Balatonfüred, (1969), pp. 24–29, (Bolyai János Matematikai Társulat, Budapest, (1970)).

[141] Roberts, F. S., *Graph theory and its applications to problems of so-ciety*, NFS-CBMS Monograph no. 29, SIAM Publications, Philadelphia, PA. 1978.

[142] Robertson, N. and P. D. Seymour, Graph minors—A survey. In I. An-derson, editor, *Surveys in Combinatorics*, Cambridge Univ. Press, Cambridge, 1985, pp. 153–171.

[143] Robertson, N. and P. D. Seymour, Graph minors. II: algorithmic as-pects of treewidth, *J. Algorithms* **7**, (1986), pp. 309–322.

[144] Robertson, N. and P. D. Seymour, Graph minors V: excluding a planar graph. *J. Comb. Theory, Series B* **41**, (1986), pp. 92–114.

[145] Robertson, N. and P. D. Seymour, Graph minors. X: obstructions to tree-decompositions, *J. Comb. Theory, Series B* **52**, (1991), pp. 153–190.

[146] Romani, F., Shortest-path problem is not harder than matrix multi-plication, *Information Processing Letters* **11**, (1980), pp. 134–136.

[147] Rose, D. J., On simple characterizations of k-trees, *Discrete Math.* **7**, (1974), pp. 317–322.

[148] Rose, D. J. and R. E. Tarjan, Algorithmic aspects of vertex elimination, *Proceedings 7th Annual ACM Symposium on Theory of Computing*, (1975), pp. 245–254.

[149] Rose, D. J., R. E. Tarjan, and G. S. Lueker, Algorithmic aspects of vertex elimination on graphs, *SIAM Journal on Computing* **5**, (1976), pp. 266–283.

[150] Scheffler, P., *Die Baumweite von Graphen als eine Maßfür die Kompliziertheit algorithmischer Probleme*, Ph.D. thesis, Akademie der Wissenschaften der DDR, Berlin, (1989).

[151] Scheinerman, E. R., Random interval graphs, *Combinatorica* **8**, (1988), pp. 357–371.

[152] Sen, A., H. Deng and S. Guha, On a graph partition problem with application to VLSI layout, *Information Processing Letters* **43**, (1992), pp. 87–94.

[153] Seymour, P. and R. Thomas, Call routing, rat catching, and planar branch width, *DIMACS Workshop—Planar Graphs: Structures and Algorithms*, Center for Discrete Mathematics & Theorical Computer Science, Nov 1991.

[154] Simon, K., *Effiziente Algorithmen für perfekte Graphen*, Leitfäden und Monographien der Informatik, Eidg. Technische Hogschule Zürich (1992).

[155] Spinrad, J., On comparability and permutation graphs, *SIAM J. Comp.* **14**, (1985), pp. 658–670.

[156] Spinrad, J., Doubly lexical ordering of dense matrices, manuscript, Department of Computer Science, Vanderbilt University, Nashville, TN, 1988.

[157] Spinrad, J., A. Brandstädt, L. Stewart, Bipartite permutation graphs, *Discrete Applied Mathematics* **18**, (1987), pp. 279–292.

[158] Strassen, V., Algebraic Complexity Theory. In: J. van Leeuwen (ed.) *Handbook of Theoretical Computer Science, A: Algorithms and Complexity*, Elsevier Science Publ., Amsterdam, 1990, pp. 633–672.

[159] Sundaram, R., K. Sher Singh and C. Pandu Rangan, Treewidth of circular arc graphs. To appear in *SIAM J. Disc. Math.*

[160] Supowit, K, J., Finding a maximum planar subset of a set of nets in a channel, *IEEE Trans. Computer Aided Design* **6**, (1987), pp. 93–94.

[161] Supowit, K. J., Decomposing a set of points into chains, with applications to permutation and circle graphs, *Information Processing Letters* **21**, (1985), pp. 249–252.

[162] Tarjan, R. E., Decomposition by clique separators, *Discrete Mathematics* **55**, (1985), pp. 221–232.

[163] Tellegen, B. D. H., Geometrical configurations and duality of electrical networks, *Philips Technical Rev.* **5**, (1940), pp. 324–330.

[164] Thomason, A., An extremal function for contractions of graphs, *Math. Proc. Cambr. Phil. Soc.* **95**, (1984), pp. 261–265.

[165] Trotter, W. T., *Combinatorics and Partially Ordered sets: Dimension Theory*, The John Hopkins University Press, Baltimore, Maryland, 1992.

[166] Tsukiyama, S., Algorithm for generating all maximal independent sets, *Electronics & Communications* **59** (1976), pp. 1–8.

[167] Véga, W. F. de la, On the bandwidth of random graphs, *Ann. Disc. Math.* **17**, (1983), pp. 633–638.

[168] Véga, W. F. de la and Y. Manoussakis, Grids in random graphs, *Random Structures and Algorithms* **5**, (1994), pp. 329–336.

[169] Wagner, K., Über eine Eigenschaft der ebenen Komplexe, *Math. Ann.* **114**, (1937), pp. 570–590.

[170] Wagner, K., Monotonic coverings of finite sets, *Journal of Information Processing and Cybernetics*, EIK, **20**, (1984), pp. 633–639.

[171] Wald, J. A. and C. J. Colbourn, Steiner trees, partial 2-trees and minimum IFI networks, *Networks* **13**, (1983), pp. 159–167.

[172] Walter, J. R., Representations of chordal graphs as subtrees of a tree, *J. Graph Theory* **2**, (1978), pp. 265–267.

[173] Whitesides, S. H., An Algorithm for finding clique cut-sets, *Information Processing Letters* **12**, (1981), pp. 31–32.

[174] Yannakakis, M., Computing the minimum fill-in is NP-complete, *SIAM J. Alg. Disc. Meth.* **2**, (1981), pp. 77–79.

Index

Springer-Verlag
and the Environment

We at Springer-Verlag firmly believe that an international science publisher has a special obligation to the environment, and our corporate policies consistently reflect this conviction.

We also expect our business partners – paper mills, printers, packaging manufacturers, etc. – to commit themselves to using environmentally friendly materials and production processes.

The paper in this book is made from low- or no-chlorine pulp and is acid free, in conformance with international standards for paper permanency.

Printing: Weihert-Druck GmbH, Darmstadt
Binding: Theo Gansert Buchbinderei GmbH, Weinheim

Lecture Notes in Computer Science

For information about Vols. 1–762
please contact your bookseller or Springer-Verlag

Vol. 763: F. Pichler, R. Moreno Díaz (Eds.), Computer Aided Systems Theory – EUROCAST '93. Proceedings, 1993. IX, 451 pages. 1994.

Vol. 764: G. Wagner, Vivid Logic. XII, 148 pages. 1994. (Subseries LNAI).

Vol. 765: T. Helleseth (Ed.), Advances in Cryptology – EUROCRYPT '93. Proceedings, 1993. X, 467 pages. 1994.

Vol. 766: P. R. Van Loocke, The Dynamics of Concepts. XI, 340 pages. 1994. (Subseries LNAI).

Vol. 767: M. Gogolla, An Extended Entity-Relationship Model. X, 136 pages. 1994.

Vol. 768: U. Banerjee, D. Gelernter, A. Nicolau, D. Padua (Eds.), Languages and Compilers for Parallel Computing. Proceedings, 1993. XI, 655 pages. 1994.

Vol. 769: J. L. Nazareth, The Newton-Cauchy Framework. XII, 101 pages. 1994.

Vol. 770: P. Haddawy (Representing Plans Under Uncertainty. X, 129 pages. 1994. (Subseries LNAI).

Vol. 771: G. Tomas, C. W. Ueberhuber, Visualization of Scientific Parallel Programs. XI, 310 pages. 1994.

Vol. 772: B. C. Warboys (Ed.),Software Process Technology. Proceedings, 1994. IX, 275 pages. 1994.

Vol. 773: D. R. Stinson (Ed.), Advances in Cryptology – CRYPTO '93. Proceedings, 1993. X, 492 pages. 1994.

Vol. 774: M. Banâtre, P. A. Lee (Eds.), Hardware and Software Architectures for Fault Tolerance. XIII, 311 pages. 1994.

Vol. 775: P. Enjalbert, E. W. Mayr, K. W. Wagner (Eds.), STACS 94. Proceedings, 1994. XIV, 782 pages. 1994.

Vol. 776: H. J. Schneider, H. Ehrig (Eds.), Graph Transformations in Computer Science. Proceedings, 1993. VIII, 395 pages. 1994.

Vol. 777: K. von Luck, H. Marburger (Eds.), Management and Processing of Complex Data Structures. Proceedings, 1994. VII, 220 pages. 1994.

Vol. 778: M. Bonuccelli, P. Crescenzi, R. Petreschi (Eds.), Algorithms and Complexity. Proceedings, 1994. VIII, 222 pages. 1994.

Vol. 779: M. Jarke, J. Bubenko, K. Jeffery (Eds.), Advances in Database Technology — EDBT '94. Proceedings, 1994. XII, 406 pages. 1994.

Vol. 780: J. J. Joyce, C.-J. H. Seger (Eds.), Higher Order Logic Theorem Proving and Its Applications. Proceedings, 1993. X, 518 pages. 1994.

Vol. 781: G. Cohen, S. Litsyn, A. Lobstein, G. Zémor (Eds.), Algebraic Coding. Proceedings, 1993. XII, 326 pages. 1994.

Vol. 782: J. Gutknecht (Ed.), Programming Languages and System Architectures. Proceedings, 1994. X, 344 pages. 1994.

Vol. 783: C. G. Günther (Ed.), Mobile Communications. Proceedings, 1994. XVI, 564 pages. 1994.

Vol. 784: F. Bergadano, L. De Raedt (Eds.), Machine Learning: ECML-94. Proceedings, 1994. XI, 439 pages. 1994. (Subseries LNAI).

Vol. 785: H. Ehrig, F. Orejas (Eds.), Recent Trends in Data Type Specification. Proceedings, 1992. VIII, 350 pages. 1994.

Vol. 786: P. A. Fritzson (Ed.), Compiler Construction. Proceedings, 1994. XI, 451 pages. 1994.

Vol. 787: S. Tison (Ed.), Trees in Algebra and Programming – CAAP '94. Proceedings, 1994. X, 351 pages. 1994.

Vol. 788: D. Sannella (Ed.), Programming Languages and Systems – ESOP '94. Proceedings, 1994. VIII, 516 pages. 1994.

Vol. 789: M. Hagiya, J. C. Mitchell (Eds.), Theoretical Aspects of Computer Software. Proceedings, 1994. XI, 887 pages. 1994.

Vol. 790: J. van Leeuwen (Ed.), Graph-Theoretic Concepts in Computer Science. Proceedings, 1993. IX, 431 pages. 1994.

Vol. 791: R. Guerraoui, O. Nierstrasz, M. Riveill (Eds.), Object-Based Distributed Programming. Proceedings, 1993. VII, 262 pages. 1994.

Vol. 792: N. D. Jones, M. Hagiya, M. Sato (Eds.), Logic, Language and Computation. XII, 269 pages. 1994.

Vol. 793: T. A. Gulliver, N. P. Secord (Eds.), Information Theory and Applications. Proceedings, 1993. XI, 394 pages. 1994.

Vol. 794: G. Haring, G. Kotsis (Eds.), Computer Performance Evaluation. Proceedings, 1994. X, 464 pages. 1994.

Vol. 795: W. A. Hunt, Jr., FM8501: A Verified Microprocessor. XIII, 333 pages. 1994.

Vol. 796: W. Gentzsch, U. Harms (Eds.), High-Performance Computing and Networking. Proceedings, 1994, Vol. I. XXI, 453 pages. 1994.

Vol. 797: W. Gentzsch, U. Harms (Eds.), High-Performance Computing and Networking. Proceedings, 1994, Vol. II. XXII, 519 pages. 1994.

Vol. 798: R. Dyckhoff (Ed.), Extensions of Logic Programming. Proceedings, 1993. VIII, 362 pages. 1994.

Vol. 799: M. P. Singh, Multiagent Systems. XXIII, 168 pages. 1994. (Subseries LNAI).

Vol. 800: J.-O. Eklundh (Ed.), Computer Vision – ECCV '94. Proceedings 1994, Vol. I. XVIII, 603 pages. 1994.

Vol. 801: J.-O. Eklundh (Ed.), Computer Vision – ECCV '94. Proceedings 1994, Vol. II. XV, 485 pages. 1994.

Vol. 802: S. Brookes, M. Main, A. Melton, M. Mislove, D. Schmidt (Eds.), Mathematical Foundations of Programming Semantics. Proceedings, 1993. IX, 647 pages. 1994.

Vol. 803: J. W. de Bakker, W.-P. de Roever, G. Rozenberg (Eds.), A Decade of Concurrency. Proceedings, 1993. VII, 683 pages. 1994.

Vol. 804: D. Hernández, Qualitative Representation of Spatial Knowledge. IX, 202 pages. 1994. (Subseries LNAI).

Vol. 805: M. Cosnard, A. Ferreira, J. Peters (Eds.), Parallel and Distributed Computing. Proceedings, 1994. X, 280 pages. 1994.

Vol. 806: H. Barendregt, T. Nipkow (Eds.), Types for Proofs and Programs. VIII, 383 pages. 1994.

Vol. 807: M. Crochemore, D. Gusfield (Eds.), Combinatorial Pattern Matching. Proceedings, 1994. VIII, 326 pages. 1994.

Vol. 808: M. Masuch, L. Pólos (Eds.), Knowledge Representation and Reasoning Under Uncertainty. VII, 237 pages. 1994. (Subseries LNAI).

Vol. 809: R. Anderson (Ed.), Fast Software Encryption. Proceedings, 1993. IX, 223 pages. 1994.

Vol. 810: G. Lakemeyer, B. Nebel (Eds.), Foundations of Knowledge Representation and Reasoning. VIII, 355 pages. 1994. (Subseries LNAI).

Vol. 811: G. Wijers, S. Brinkkemper, T. Wasserman (Eds.), Advanced Information Systems Engineering. Proceedings, 1994. XI, 420 pages. 1994.

Vol. 812: J. Karhumäki, H. Maurer, G. Rozenberg (Eds.), Results and Trends in Theoretical Computer Science. Proceedings, 1994. X, 445 pages. 1994.

Vol. 813: A. Nerode, Yu. N. Matiyasevich (Eds.), Logical Foundations of Computer Science. Proceedings, 1994. IX, 392 pages. 1994.

Vol. 814: A. Bundy (Ed.), Automated Deduction—CADE-12. Proceedings, 1994. XVI, 848 pages. 1994. (Subseries LNAI).

Vol. 815: R. Valette (Ed.), Application and Theory of Petri Nets 1994. Proceedings. IX, 587 pages. 1994.

Vol. 816: J. Heering, K. Meinke, B. Möller, T. Nipkow (Eds.), Higher-Order Algebra, Logic, and Term Rewriting. Proceedings, 1993. VII, 344 pages. 1994.

Vol. 817: C. Halatsis, D. Maritsas, G. Philokyprou, S. Theodoridis (Eds.), PARLE '94. Parallel Architectures and Languages Europe. Proceedings, 1994. XV, 837 pages. 1994.

Vol. 818: D. L. Dill (Ed.), Computer Aided Verification. Proceedings, 1994. IX, 480 pages. 1994.

Vol. 819: W. Litwin, T. Risch (Eds.), Applications of Databases. Proceedings, 1994. XII, 471 pages. 1994.

Vol. 820: S. Abiteboul, E. Shamir (Eds.), Automata, Languages and Programming. Proceedings, 1994. XIII, 644 pages. 1994.

Vol. 821: M. Tokoro, R. Pareschi (Eds.), Object-Oriented Programming. Proceedings, 1994. XI, 535 pages. 1994.

Vol. 822: F. Pfenning (Ed.), Logic Programming and Automated Reasoning. Proceedings, 1994. X, 345 pages. 1994. (Subseries LNAI).

Vol. 823: R. A. Elmasri, V. Kouramajian, B. Thalheim (Eds.), Entity-Relationship Approach — ER '93. Proceedings, 1993. X, 531 pages. 1994.

Vol. 824: E. M. Schmidt, S. Skyum (Eds.), Algorithm Theory – SWAT '94. Proceedings. IX, 383 pages. 1994.

Vol. 825: J. L. Mundy, A. Zisserman, D. Forsyth (Eds.), Applications of Invariance in Computer Vision. Proceedings, 1993. IX, 510 pages. 1994.

Vol. 826: D. S. Bowers (Ed.), Directions in Databases. Proceedings, 1994. X, 234 pages. 1994.

Vol. 827: D. M. Gabbay, H. J. Ohlbach (Eds.), Temporal Logic. Proceedings, 1994. XI, 546 pages. 1994. (Subseries LNAI).

Vol. 828: L. C. Paulson, Isabelle. XVII, 321 pages. 1994.

Vol. 829: A. Chmora, S. B. Wicker (Eds.), Error Control, Cryptology, and Speech Compression. Proceedings, 1993. VIII, 121 pages. 1994.

Vol. 830: C. Castelfranchi, E. Werner (Eds.), Artificial Social Systems. Proceedings, 1992. XVIII, 337 pages. 1994. (Subseries LNAI).

Vol. 831: V. Bouchitté, M. Morvan (Eds.), Orders, Algorithms, and Applications. Proceedings, 1994. IX, 204 pages. 1994.

Vol. 832: E. Börger, Y. Gurevich, K. Meinke (Eds.), Computer Science Logic. Proceedings, 1993. VIII, 336 pages. 1994.

Vol. 833: D. Driankov, P. W. Eklund, A. Ralescu (Eds.), Fuzzy Logic and Fuzzy Control. Proceedings, 1991. XII, 157 pages. 1994. (Subseries LNAI).

Vol. 834: D.-Z. Du, X.-S. Zhang (Eds.), Algorithms and Computation. Proceedings, 1994. XIII, 687 pages. 1994.

Vol. 835: W. M. Tepfenhart, J. P. Dick, J. F. Sowa (Eds.), Conceptual Structures: Current Practices. Proceedings, 1994. VIII, 331 pages. 1994. (Subseries LNAI).

Vol. 836: B. Jonsson, J. Parrow (Eds.), CONCUR '94: Concurrency Theory. Proceedings, 1994. IX, 529 pages. 1994.

Vol. 837: S. Wess, K.-D. Althoff, M. M. Richter (Eds.), Topics in Case-Based Reasoning. Proceedings, 1993. IX, 471 pages. 1994. (Subseries LNAI).

Vol. 838: C. MacNish, D. Pearce, L. Moniz Pereira (Eds.), Logics in AI. Proceedings, 1994. IX, 413 pages. 1994. (Subseries LNAI).

Vol. 839: Y. G. Desmedt (Ed.), Advances in Cryptology - CRYPTO '94. Proceedings, 1994. XII, 439 pages. 1994.

Vol. 840: G. Reinelt, The Traveling Salesman. VIII, 223 pages. 1994.

Vol. 841: I. Prívara, B. Rovan, P. Ružička (Eds.), Mathematical Foundations of Computer Science 1994. Proceedings, 1994. X, 628 pages. 1994.

Vol. 842: T. Kloks, Treewidth. IX, 209 pages. 1994.

Vol. 843: A. Szepietowski, Turing Machines with Sublogarithmic Space. VIII, 115 pages. 1994.